BEHAVIORAL VARIATION

Case Study of a Malagasy Lemur

Alison F. Richard

LEWISBURG
BUCKNELL UNIVERSITY PRESS
LONDON: ASSOCIATED UNIVERSITY PRESSES

© 1978 by Associated University Presses, Inc.

Associated University Presses, Inc.
Cranbury, New Jersey 08512

Associated University Presses
Magdalen House
136-148 Tooley Street
London SE1 2TT, England

Library of Congress Cataloging in Publication Data

Richard, Alison F.
    Behavioral variation.

    (The Primates)
    Bibliography: p.
    Includes index.
    1. Verreaux's sifaka — Behavior.   2. Social behavior in animals.
    3. Mammals — Behavior.   4. Mammals — Madagascar.   I. Title
    QL737.P945R5   599'.81   76-19837
    ISBN 0-8387-1965-1

For my parents

Contents

Contents 9

List of Tables

11

List of Figures

13

List of Plates

Acknowledgments

Adequate expression of my appreciation to the many people who have contributed in all kinds of ways to this book would require a volume of its own. Here, I can only acknowledge special debts — to my adviser, John Napier, whose insight, experience and sense of humor did much to make it all possible; to Alison Jolly, but for whom it would never have been done at all, and to David Pilbeam who, as teacher and colleague, was and is my primary source of inspiration.

I would also like to express special gratitude to the Malagasy Government for permitting this study to be made, and to the many people in Madagascar whose advice, support, and hospitality played an important role in its execution. My thanks go to the staff of the Direction de la Recherche Scientifique et Technique and, particularly, to the Director, Dr. E. Rakotomarie, to Georges Randrianasolo, Director of the National Zoological Park, to the staff of the Forestry Department, especially M. Andriamampianina, to Folo Emmanuel, and to Guy Ramanantsoa. I suspect that my field assistants, Ranarivelo Marly and K. Augustin, will never recognize just how much I owe them: their perseverance, good humor, moral support and, on more than one occasion, digging abilities, were forever saving me and/or the landrover from imminent disaster. Thanks are also due to the De Heaulme family, whose foresight provided the reserve at Berenty and whose generous hospitality permitted me to work there. At various points on my travels a noble band of people regularly fed and watered me, and I would like to thank them too: the Ulrichs, the Ashcrofts, the Clarks, the Warders, Ro Booth, Wil Wilson, and the R.A.F. detachment in Majunga.

I am most grateful to Dr. J. J. Petter, who gave me his time and his advice, and freely shared his wide experience of Madagascar with me. Bob and Linda Sussman are valued friends and colleagues both in and

out of the field, and to them too I offer special thanks. I also appreciate
the many useful criticisms and comments that have been offered me on
various aspects of the study over the last six years: in England, my
thanks go to Tim Clutton-Brock, John Crook, Robert Hinde, Peter
Lattin, and Bob Martin; in America, to Lee McGeorge, Serge Goldstein,
Glenn Hausfater, Jeff Laitman, David Post, Jay Russell, Ian Tattersall,
and Kathy Wolf; and in the forest, to Tom Struhsaker. To Susan Klein
go my delighted thanks for the drawings for the book's jacket.

The study was generously supported by a Royal Society Leverhulme
Award, the Explorers' Club of America, the Boise Fund, the Society of
the Sigma Xi, a NATO Overseas Studentship, the John Spedan Lewis
Trust Fund for the Advancement of Science, the Central Research
Fund of London University, and National Science Foundation Grant
No. 46014 B043188.

I thank the University of Chicago Press for permission to reprint
from Alison Jolly, *Lemur Behavior*, 1966, Table II-1, p. 26. I also
thank S. Karger AG, Basel, for permission to reprint from Alison
Jolly, *Folia primatologica* 17, 1972, Table II, p. 340.

Finally, my thanks and appreciation to Bob Dewar, who arrived on
the scene in time for the prolonged final throes of this book. I can
think of no one I would rather have final throes around.

Alison F. Richard

BEHAVIORAL VARIATION

1
Introduction

en somme, ceux sont des animaux peu actifs, peu remuants, peu intelligents.

— Milne-Edwards and Grandidier (1886)

The remaining forests of Madagascar still harbor the vestiges of a unique and little studied fauna. Probably the best known among members of the six mammalian orders represented there are the lemurs, variously classified as the Infraorder Lemuriformes in the Suborder Prosimii (Simpson, 1945), or the Suborder Lemuroidea in the Grade Strepsirhini (Hill, 1953).

Between April 1970 and September 1971, an intensive study was made in Madagascar of one of the largest and most widely distributed of the lemurs, *Propithecus verreauxi*. Four groups were selected to be the focus of observations. Two of these groups, belonging to the *coquereli* subspecies (Plate I), lived in rich, mixed deciduous, and evergreen forest in the northwest of the island. The other two, of the *verreauxi* subspecies (Plate II), lived in parched Didiereaceae forest, or "forêt épineuse," in the extreme south. Data were collected on various aspects of the ecology and social organization of these groups: diet and feeding behavior were studied in relation to patterns of ranging and home-range utilization. Seasonal changes in social relationships both within and between groups were investigated, particularly with reference to behavioral changes in the mating season. In order to compare systematically differences in the habitat at each study area, vegetation in both areas was analyzed and records kept of some climatic variables.

21

PLATE I. The northwestern subspecies, *Propithecus verreauxi coquereli*

Between May and August 1974 a follow-up study was carried out, which included further observations on one of the four groups studied in 1970/71. These were made to provide data for comparison with those collected in the earlier study. The aim of this comparison was to assess the long-term stability of ranging and feeding patterns, and of social organization. Broader surveys and censuses were also carried out in 1974 in order to widen the significance of the focal study.

Chapter 1 of this book introduces the study with a description of its aims and of the Malagasy lemurs in general, *Propithecus verreauxi* in particular. Chapter 2 describes the methods used and the ecology of the two study areas. Chapters 3, 4, and 5 present the results of censuses, and of observations in the study areas, with emphasis on local, regional, seasonal, and long-term fluctuations in behaviors and environmental variables. Chapters 6, 7, and 8 contain a discussion of the results. An attempt is made to integrate them, and to place them in a broader — and in parts highly speculative — context.

PLATE II. The southern subspecies, *Propithecus verreauxi verreauxi*

1. *Propithecus verreauxi:* A Malagasy Prosimian

Throughout most of the Mesozoic, Madagascar formed part of the African mainland but, toward the end of this period, a large land mass began to drift away from the east coast of Africa. Today, Madagascar lies about 400 km off the Mozambique coast (Fig. 1). About 1500 km long and stretching from 12° to 26°S., it is one of the largest islands in the world. A combination of climate, topography and, latterly, human interference has produced a wide range of environment on this island so that in many respects it resembles a microcosmal continent. Wet littoral forests on the east coast give way to rain forests on the steep east coast escarpment. The central plateau is now largely denuded of trees and in many areas sterile, its soil eroded by prolonged exposure to sun, wind, and rain. The west of the island is covered by a dry woodland-savanna which gives way in the wetter north to mixed deciduous and evergreen forest and in the arid south to cactuslike Didierea forest and scrub or bush.

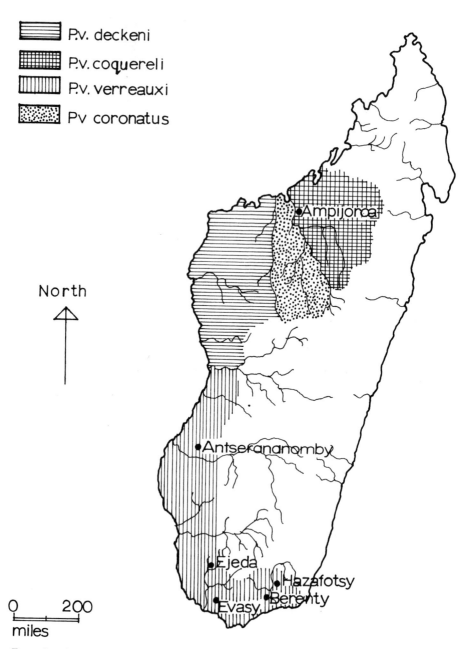

FIG. 1. The approximate distribution of the four subspecies, *P. v. coquereli, P. v. deckeni, P. v. coronatus* and *P. v. verreauxi*, and the location of study areas and forests in which surveys were made.

It is now thought that at the time of its separation from the mainland about 50 million years ago, Madagascar was populated by a "handful of ancestral mammals"; these animals, some of them ancestors of living lemurs, have evolved in isolation on the island ever since (Martin 1972). During this time, unlike the prosimians of mainland Africa, they met with little predation and no competition from the Anthropoid primates. As a result, the lemurs radiated and specialized to occupy a wide diversity of niches, many of which are occupied by the Anthropoidea throughout the rest of the Old World. As recently as two thousand years ago there were at least thirty-three species living on the island, but only nineteen survived the impact of human colonization over the past fifteen hundred years (Martin, 1972a; Walker, 1967). Little is known of the immediate causes underlying the rapid extinction of so many species, but it is likely that they succumbed to the pressures imposed by hunting and progressive deforestation (Richard and Sussman, 1975).

Propithecus verreauxi, together with *P. diadema, Indri indri,* and *Avahi laniger*, constitutes the Family Indriidae within the Lemuriformes (Simpson, 1945). *P. verreauxi* is found in forested areas of the northwest, west, and south of the island (Fig. 1). With the exception of the mouse lemur, *Microcebus murinus*, it is the most widely distributed of the lemurs; its range includes rich, mixed deciduous and evergreen forest, tamarind-dominated gallery forest, and the semi-arid thorny forest of the south. Four subspecies are recognized: *P. v. verreauxi, P. v. deckeni, P. v. coquereli,* and *P. v. coronatus.* It is likely that this differentiation is of recent origin, occurring as a consequence of the deforestation of the central plateau in Madagascar; until that time, the presence of forest at the head waters of the large rivers of western Madagascar meant that these rivers did not constitute an effective barrier between populations. Today, it is thought that the Tsiribihina River separates *verreauxi* and *deckeni* (but confirmation is needed) and that the Betsiboka River forms a barrier between *deckeni* and *coquereli.* Further surveys are needed to establish the location and nature of the boundary between *coronatus* and *deckeni.* One anomaly should be noted here: *verreauxi* and *coquereli* resemble each other more morphologically than does either the geographically intermediate *deckeni* or *coronatus* populations (Martin 1972a).

Throughout its range, *P. verreauxi* is found living sympatrically with other lemur species, including both diurnal and nocturnal forms. In the northwest, for example, it is sympatric with *Lemur fulvus, L. mongoz, Avahi laniger, Cheirogaleus medius, Lepilemur mustelinus,* and *Microcebus murinus.* In 1966 Jolly commented that lemur ". . . niches may largely overlap in seasons of abundance and be differentiated only in times of scarcity." However, distinct niches occupied by different species have been identified in subsequent ecological studies (e.g., Charles-Dominique and Hladik 1971; Martin 1972b; Petter et al. 1971; Pollock 1975a, 1975b; Sussman 1974).

2. Previous Studies of *Propithecus verreauxi*

Sieur Etienne de Flacourt's description of *P. verreauxi* in 1661 was the first written account of the species: "Il y a encor vne espèce de gnenuche blanche, qui a vn chaperon tanné, et qui se tient le plus souuent sur les pieds de derrier. . . . " Flacourt subsequently died of gout, and no more was heard of the species until Bennett gave it the generic name *Propithecus* in 1832. *P. verreauxi* was scientifically named as a species by A. Grandidier in 1867, and the four subspecies were named between 1867 and 1931.

The nineteenth century was a period of collection and of descriptive and anatomical studies, which reached their zenith in Milne-Edwards's and Grandidier's detailed account of all the lemurs in the "Histoire naturelle des mammifères: Histoire physique, naturelle, et politique de Madagascar" (1890-1896). These studies have recently been surveyed by Hill (1953).

Behavioral data collected up until 1962 were very limited but have been shown by further research to have been substantially accurate. The first reference to lemur behavior was made by Shaw (1879), but the only extensive description of *P. verreauxi*'s behavior was given by Rand (1935). He recorded *P. v. deckeni* as living in groups of up to nine, and described their locomotion: ". . . when the animals were travelling slowly, they sometimes took a few waddling steps. At no time did they travel on all fours. . . . "

Petter (1962a) made the first major survey of lemur ecology and behavior, including an account of *P. v. coquereli*. Salient features of this subspecies' natural history, according to Petter, were as follows: animals generally lived in "family" groups containing one adult male, one adult female, and one or two offspring. Each group maintained a territory. Little social interaction occurred other than between mother and infant, and mating occurred seasonally only — sometime between January and March. There was seasonal variation in animals' daily activity patterns, though it was characterized overall by morning, midday, and late-afternoon peaks of feeding. Animals ate buds, leaf shoot, leaves, bark, and fruit.

Petter's account provided the basis for Jolly's (1966) subsequent, more detailed study of *P. v. verreauxi* and for my own study of both subspecies. Jolly's observations are referred to and used as comparative evidence throughout this book, and thus no summary of her findings is presented here.

Although studies of various aspects of prosimian behavior in captivity have been made (Andrew 1963a, 1963b; Bishop 1962; Buettner-Janusch and Andrew 1962; Evans and Goy 1968; Jolly 1964a, 1964b; Klopfer 1972; Klopfer and Klopfer 1970; Petter-Rousseaux 1962, 1964), none of them related specifically to *P. verreauxi*. This is probably due to the difficulty found in keeping animals alive in captivity. At present, the only breeding colony in existence outside Madagascar is at the Duke Primate Facility in North Carolina.

3. Aims of This Study

The study had three primary objectives. The first was to assemble detailed ecological and behavioral data on a prosimian species, in order to provide a broader overview of the whole spectrum of primate adaptations and, possibly, some insight into ancestral primate patterns of behavior (see also Martin 1972a). While the literature on the Haplorhini has proliferated during the last fifteen years, the prosimians have been much less studied; at the outset of this research, Petter's surveys (1962a, 1962b, 1962c, 1965) and Jolly's field study were the only major published works on the behavior and ecology of any prosimian species in the wild. (The situation has since been much improved by the publication of two major collections of papers on prosimian research [Martin, Walker and Doyle 1974; Tattersall and Sussman 1975].)

A second aim of the study was to investigate the flexibility in time and space of the social organization of a prosimian species by comparing over time groups from populations living in widely contrasting habitats. The discovery of extensive regional variations in social organization has demonstrated the limited applicability of the concept of "species-specific" behavior to several Old World primate species (*Papio anubis* − Hall and DeVore 1965; Rowell 1966; *Presbytis entellus* − Jay 1965; Ripley 1967; Sugiyama 1967; Yoshiba 1968; *Cercopithecus aethiops* − Gartlan and Brain 1968; Struhsaker 1967). No such investigation of any prosimian species had been made prior to this study.

The study's third objective was to use a comparative approach in an effort to understand more clearly the processes by which ecology may influence social organization. Attempts have been made to produce a classification of primate social organizations, and to correlate variations in social organization with ecology (Crook 1970; Crook and Gartlan 1966; DeVore 1963; Eisenberg et al 1972; Hall 1965a). These efforts have not been altogether successful, partly for lack of detailed information on the behavior and ecology of a wide range of primate species. Since there are now extensive data on Old World and New World leaf-eating monkeys (Bernstein 1968; Chivers 1969; Clutton-Brock 1972; Marler 1969; Oates 1974; Poirier 1969; Richard 1970; Ripley 1970; Struhsaker 1975; Struhsaker and Oates 1975; Yoshiba 1968), this study set out to provide comparative material on the ecology of a prosimian species, a species whose diet consisted largely of leaves and shoots.

2
Description of
Methods and Study Areas

A. METHODS

1. Broad Census Techniques

In all forests surveyed in 1970/71 and 1974, animals were located from already existing trails and then followed until all animals present had been counted, sexed, and assigned to an age class. When three counts of a given group had been made on separate occasions, and all had provided the same information on group size and composition, they were considered to be "good" counts. (This assumed that apparent changes in group size were due to observer error rather than to real changes within the group.) At least one marker animal in each group could always be distinguished, thus avoiding the possibility of censusing one group twice.

2. Preparation of Study Areas

a) *Selection*

There were three principal considerations involved in the selection of the two main study areas:

i) Abundance of *P. verreauxi*

For the investigation of intergroup relationships, it was important

29

that the study groups should be part of a continuously distributed population rather than isolated units. Further, the presence of isolated groups was likely to be a result of persistent hunting, and remaining groups would probably be wary and difficult to habituate. The incidence of hunting varied enormously from region to region, depending on the traditions of people living in the area.

ii) Condition of forest

Most of the more accessible forests of Madagascar have been disturbed and transformed by man, through wide-scale timbering, slash-and-burn cultivation, cutting for firewood, and grazing cattle (Richard and Sussman 1975). Study areas not subject to such interference were sought, so that animals could be observed living in habitats whose characteristics and stability were not affected by man.

iii) Accessibility of study areas

For a valid comparison of the behavior and ecology of animals living in two contrasting habitats, prolonged observations had to be made in both areas in the wet and dry season. Thus, round-the-year accessibility was vital.

Largely because of this prerequisite of accessibility, both the study areas finally chosen were, or had been, subject to some direct interference by man. However, *P. verreauxi* was abundant in both, there was hunting in neither, and the degree of interference appeared to be quite limited.

b) *Location*

The northern study area 16° 35′ S. and 46°82′ E.) was situated in mixed deciduous and evergreen forest (Plate III) on a flat hilltop in the forestry reserve at Ampijoroa, in a region known as the Ankarafantsika (Fig. 1). It was about 100 kms from the west coast of the island, and 2 kms from Lac Ampijoroa. No timbering was permitted in the reserve, and livestock were never seen in the study area although cattle were seen grazing in other parts of the forest. *P. v. coquereli* was sympatric with six other primate species: *Lemur fulvus fulvus, L. mongoz, Microcebus murinus, Cheirogaleus medius, Lepilemur mustelinus,* and *Avahi laniger.*

The southern study area (24° 85′ S and 46° 50′ E) was situated in semi-arid didierea forest (Plate IV) 1 km from Hazafotsy, around 1,000 kms south of Ampijoroa (Fig. 1). Hazafotsy is about 100 kms northwest of Fort Dauphin and 60 kms from the south coast. It lies on the north-west boundary of Reserve Nationale No. 11, and the study area was inside the reserve. That sector of the reserve is flat, but it is enclosed by spurs of the Anosy mountain chain only 4 kms to the north, south,

PLATE III. The northern study area at Ampijoroa

PLATE IV. The southern study area at Hazafotsy

and east. The forest in the study area had not been burned or felled in recent times, but cattle and goats grazed in it throughout the year, and dead wood was cut up by the villagers for firewood. Only two primate species other than *P. v. verreauxi* were seen in the southern forest: *Microcebus murinus* and *Lepilemur mustelinus*.

c) *Gridding*

The two study areas were gridded into quadrats with sides of 50 m by trails marked at ground level. No precise estimate of the error resulting from inaccurate mapping was made, but all trails were paced and two, where the margin of error exceeded ±10 m, were recut.

The main purpose of the grid was to plot the movements of each study group and to determine how much time it spent in different parts of its home-range. The rationale for the 50 m-sided quadrat was that it represented the minimum (and thus the most accurate) unit of measurement that could be implemented. In effect, the grid divided up each group's home-range into about 36 quadrats.

3. Sampling of Vegetation

In both focal study areas, two vegetation samples were taken in each hectare quadrat delineated by the grid. Within each hectare quadrat the two samples were randomly located, following the "stratified random" sample technique described by Southwood (1966). This method ensures that the total sample is well distributed over the total area to be sampled. At the same time, the bias arising if species distributed in a uniform pattern are also sampled according to a uniform pattern is minimized by randomization within each hectare quadrat.

Each sample consisted of a circle with a measured radius of 5 m. Within each circle, all trees with a trunk diameter greater than 3 cm were counted and identified, and the trunk diameter, height and maximum spread of the tree, height of the lowest branch above the ground, and phenology were noted. Tree height, maximum spread, and the height of the lowest branch above the ground were estimated by eye. A sample of 200 trees whose height had been estimated by eye was subsequently measured. Estimates were found to be accurate within ±2 m of the measured heights.

A total of 2,619 trees was described in the northern study area, taken from 24 samples. In the south, 3,136 trees coming from 30 samples were described. The total sample covered 1.57% of the surface area of each study area. In both areas, lianas constituted an important item in the diet of *P. verreauxi*. It was not possible to make an accurate estimate of the density of lianas, but a count was made of the number of trees bearing lianas in each sample in order to provide a rough estimate of abundance.

Data on synchrony of leaf and fruit production and their presence in tree species were based on observations made on ten species in each study area. Not all the preferred food species of *P. verreauxi* were known at the outset of the study, so that some tree species were selected for sampling because they constituted a large proportion of the animals' diet during the first month of observations in each area, and others because they were abundant. Each month, observations on ten individual trees from each species were tabulated. Different trees were sampled each month. Five nonexclusive phenological categories were used:

1) leaf shoot and/or immature leaves and flush (Immature leaves were differentiated from mature leaves on the basis of their paler coloring and, sometimes, smaller size)
2) mature and/or dying leaves
3) flowers, open and /or in bud
4) fruit, green and/or ripe
5) dormant buds

A record of "present" or "absent" was entered for each of these categories, but no estimate of quantity was made. In cases where one or two desiccated fruit or dead leaves remained attached to a tree when all the rest had already dropped, "absent" was recorded in these categories.

4. Study Groups

a) *Choice of sample size in focal study*

Two neighboring groups were chosen for study in each area. This sample size was chosen to provide a compromise between breadth and depth. One aim of the study was to investigate the effects of differing habitats on the behavior of *P. verreauxi* and it was assumed that some variations in behavior might be of a subtle nature, resulting from very slight differences in habitat. If data had been in the form of surveys of several groups, they would have precluded the detailed analysis required to detect such differences. Two groups rather than one were selected in each area because extreme reduction of sample size inevitably limits the value or significance of any conclusions, and there was enough time to study both in the depth desired.

b) *Criteria for sexing and aging animals*

Sex could be determined from the appearance of the genitalia and, in adults, from the presence or absence of the throat gland, found only in males. Although subadult and juvenile males performed the rubbing action associated with throat marking, the gland did not appear as a dark strip on their throats as it did in adults.

Since the birth period was very short each year, and synchronized among the females in each region, gradations in size and, presumably, age of immature animals in each study area were clear. Further, the development of one infant from birth to the age of seven months was recorded in detail. Combining knowledge of the size of an animal at seven months with the knowledge that births occcurred at approximately yearly intervals, the ages of immature animals could be inferred on the assumption that each size gradation contained the offspring of a different birth period. It is unlikely that any stage was missed because of the small sample size since *P. verreauxi* takes only three years to reach maturity.[1] It is assumed that the categories infant, juvenile, and subadult marked this three-year maturation process.

Infants were identified as those animals which were carried part of the time by adult females. This occurred up to the age of at least seven months, but no record of development between seven months and one year was obtained. As of one year, animals moved completely independently of their mothers. Animals up to two years old that were completely independent of their mothers were called *juveniles*. Animals between two and three years old were called *subadults*. During censuses, the categories *juvenile* and *subadult* were grouped together. It is possible that growth may continue for more than three years before full adult size is reached, and that the subadult grade included animals over three years old. However, there were no clear distinctions beyond those described here.

c) *Individual recognition of animals*

Apart from sex and size/age differences, individuals were recognized by variation in features such as coat color and condition, facial scars, ear notches, broken digits and, in the south, the formation of the dark "cap line" on the head (Plate V). The absence of this cap in the northern

PLATE V. Portrait of an adult male, showing individually distinguishing features

subspecies made recognition much more difficult, although identification on a day-to-day basis was possible through temporary discolorations of fur and bare patches on the tail. All animals in the southern study groups could be identified consistently.

d) *Habituation of groups*

A group was said to be habituated when all its members would approach to within 2 m of my field assistant and me to feed (Plate VI). "Group" habituation is thus to be distinguished from the habituation

PLATE VI. A habituated animal feeds in close proximity to Augustin

of single animals within a group. Quantitative records were kept only of animals belonging to habituated groups.

When observations were first taken up in the northern study area, stalking was a practical impossibility; cover was minimal, and the forest floor was ankle deep in dead, dry leaves. Thus animals were necessarily followed quite openly, although with a minimum of sudden noise or movement; indeed, I tried to ensure that I could always be seen by them. Stoltz and Saayman (1970) used this method successfully when they were habituating *Papio ursinus*; they assumed that the flight response of the potential study groups had not been heavily reinforced through hunting by man, and felt that if contact could be maintained with the animals for sufficiently long periods of time, their continued inability to "escape," and hence the lack of reward to their flight response, would ultimately result in the extinction of that flight response.

Irrespective of the validity of the theory underlying the habituation method, it succeeded rapidly. All four groups were habituated within three-and-a-half weeks, although for the first three or four days they fled precipitously when I approached. Occasional glances at me demonstrated the animals' general awareness of my presence throughout the study, but after the habituation period they seemed to be unaffected by it.

In all four groups, habituation took place with reference to individuals rather than the presence of people generally; they showed an immediate flight response when approached by strangers.

When contact was reestablished with one of the southern study groups in 1974, animals still seemed indifferent to my presence despite the three-year interval, and within two hours of my first contact with them they were feeding within 2 m of me. Only the one-year-old juvenile (unborn in the previous study) fled; by the end of three days, this animal was habituated too.

5. Sampling Methods and Observational Techniques in Focal Study

a) *Data recording*

The northern study groups, Groups I and II, were observed during the dry season months of July and August 1970 and July 1971, and during the wet season months of October, November, and December 1970. The southern groups, Groups III and IV, were studied during the dry season months of April, May, and June 1971 and during the wet season months of January, February, and March 1971. Observations on Group III were also made in September 1970 and June 1974. During these months a total of 2500 hours was spent in observation.

One animal was followed for twelve hours each day, from 0600 h - 1800 h. Each day the age/sex class of the subject was changed and within each season observations were equally distributed between the different classes, as well as between different times of day. Three nights in each season were spent observing the animals in both study areas, and in all cases they remained inactive throughout the night. Further, animals could usually be relocated in the morning in the position they had last been seen the previous evening. Exceptions occurred only in the wet season, when animals sometimes became active before 0600 h, and continued to move and feed after 1800 h. It was too dark to record accurately at those times and animals were never observed to be active after 1900 h or before 0500 h in either season. It is thus assumed that almost all activity occurred within the daily observation period.

Data on most aspects of behavior were recorded at timed minute intervals. Seventy-two hours of data collected in this way were recorded for each group in each of the months referred to above. These seventy-two hours of observations per group per month constitute the raw data for most of the analyses presented; in these analyses, the *NUMBER OF RECORDS MADE AT MINUTE INTERVALS* that an animal spent in an activity will be referred to as *TIME SPENT* in that activity (J. Altmann 1974). Exceptions to this include (1) the distance traveled by the subject animal each day, which is based on records made at minute intervals of the distance traveled by the subject *DURING THE PRECEDING TIME SAMPLE INTERVAL*, and (2) rarely occurring behaviors. The frequency of interactions, self-grooming, scent-marking, and vocalizations was low and sampling at minute intervals tended to underestimate their importance. An additional record was thus kept, which described these activities *WHENEVER THEY OCCURRED* and involved the subject animal.

b) *Types of information recorded in focal study*

Minute-by-minute data on the subject animal were collected in 15 categories. Many of these categories contained several subdivisions. Each subdivision was numbered, so that data sheets contained only coded information in numerical form. These categories, and their subdivisions, are listed below in five groups. Further description and operational definitions of behavioral categories are given in chapters 3, 4, and 5.

    i) Identity and location of subject

       1) Identity of subject animal. All animals were individually

recognized and named. This category remained constant throughout the day.

2) Time of day at which recording was made. Unless some disturbance interrupted recordings, this record ran from 0600 h to 1800 h at minute intervals.

3) Quadrat occupied by the subject. Quadrats were identified by the surrounding marked trails.

4) The subject's height above ground. The following subdivisions were used:

> on the ground
> off the ground − 1½ m
> 1½ m - 3 m
> 3 m - 5 m
> 5 m - 7 m
> 7 m - 10 m
> 10 m - 13 m
> over 13 m

This measure was probably subject to some inaccuracy in that height was estimated by eye alone.

5) Whether the subject was in the sun or the shade.

ii) Subject's activity and related factors

1) Subject's activity. Activities were recorded in the following, mutually exclusive subdivisions:

> rest
> feed
> move
> self-groom
> interact
> scent-mark

2) Subject's posture. *P. verreauxi* has a wide range of postures associated with the whole spectrum of locomotor patterns of which the species is capable. These postures constitute a continuum that is difficult to divide up meaningfully into discrete categories. A rudimentary classification was required to investigate regional and/or interindividual differences, however, and eight basic postural categories were finally formulated. It should be stressed that the classification given below is not a full analysis of the range of postures seen; the species' locomotor "totipotentiality" (Prost 1965) was wider than suggested by such a restrictive classification:

PLATE VII. A young adult feeding in the sloth posture

       a — vertical clinging and leaping posture on a vertical support

       b — sitting on a horizontal support

       c — hanging by both arms

       d — hanging by both legs

       e — hanging by one arm and one leg

       f — lying along a horizontal support

       g — clinging by arms and legs under a network of twigs, with minimal curvature of the vertebral column

       h — hanging under a branch in a slothlike position, with the vertebral column curved, weight supported by legs and one arm (Plate VII).

3) Subject's substrate. Five different types of substrate were categorized and differentiated by eye. These related closely to the structural components of trees in both study areas:

       a — ground

       b — main (greater than 6 cm in diameter) trunk or vertical branch

       c — main (greater than 6 cm in diameter) horizontal branch

       d — branches less than 6 cm in diameter

       e — twigs less than 1½ cm in diameter

4) Food species. All food species were numbered, and the number recorded in this category if the animal was feeding. The part of the food species being eaten was described separately.

5) Distance traveled by the subject in the preceding minute. This again was measured by eye alone.

iii) Group dispersion

1) Proximity of the subject's nearest neighbor. Five categories of proximity were established

     physical contact

     1 - 2 m to nearest animal

     2 - 3 m to nearest animal

     3 - 6 m to nearest animal

     over 6 m to nearest animal

2) The identity of the subject's nearest neighbor.

iv) Vocalizations

P. verreauxi's vocal repertoire is characterized by a series of easily recognizable calls and a few "intermediate" vocalizations.[2] Seven subdivisions, representing seven classes of vocalization, were formulated, following Jolly's (1966) nomenclature except for the last:

    low growl
    sifaka
    roaring bark
    purr
    coo
    locomotor grunt
    spat

On occasion, the allocation of "intermediate" calls to these subdivisions was fairly arbitrary, but they constituted a small proportion of total calls made.

v) Environment

The weather was described in five general subdivisions while data were being collected on the animals:

    rainy
    overcast with no wind
    overcast with wind
    sunny with no wind
    sunny and windy

In addition, the temperature at 0600 h and 1800 h, and the maximum and minimum temperatures for each twenty-four hours, were recorded from a thermometer hung in the shade in the middle of each study area. Rainfall data were supplied by meteorological stations, located within 30 kms of each study area.

B. THE STUDY AREAS

1. Climate

a) *Rainfall*

Seasonal changes in rainfall were clear in the north, and extremely marked in the south. Toward the middle of October 1970, heavy rain began to fall in the north, particularly at night. Between mid-October and December, 385.4 mm of the annual total of 1,619 mm of rain fell. These months are referred to as the wet season. Between July and September, only 1.8 mm of rain fell; these months are referred to as the dry season. In the south, 613.8 mm of the annual total of 629.8 mm of rain fell during January, February, and March 1971 (the wet season);

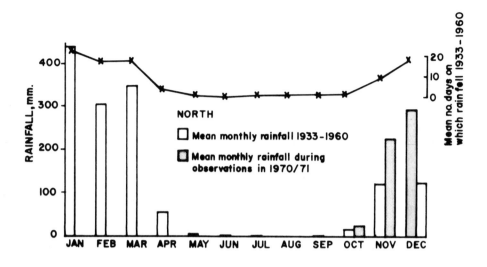

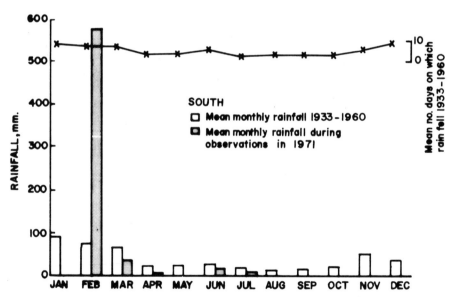

FIG. 2. Mean monthly precipitation (1933-1960), mean number of days each
month on which rain fell (1933-1960), and mean monthly precipitation
during observations in the two study areas (1970-1971).

practically no rain fell in the rest of the year, which thus constituted an extended dry season.

Figure 2 shows the mean monthly rainfall recorded at meteorological stations 30 kms from the northern study area, and 16 kms from the southern study area, between 1933 and 1960, and during the months of the field study in 1970 and 1971. It also shows the mean number of days each month on which rain fell, between 1933 and 1960. The results shown for 1970 in the north approximate those for earlier years, both in terms of the amount and seasonal distribution of rain. In the south, results are similar to those for 1970 (Charles-Dominique and Hladik 1971) but differ sharply from those for previous years: heavy rain, most of which fell during the three cyclones that swept the region during January and February (the results for January were destroyed when the meteorological station was flooded), was followed by almost total drought. In the earlier years, the general pattern was of slight precipitation throughout the year, with a peak in January.

b) *Temperature*

Daily fluctuations in temperature were considerable in both study areas, particularly during the dry season. Figure 3 shows the mean monthly maxima and minima based on readings taken every twenty-four hours. In the north, maximum and minimum temperatures of 39.5°C and 18°C were recorded during the wet season, and 31.5°C and 14°C during the dry season. In 1971 in the south, wet season maxima and minima of 39°C and 16°C were recorded, and dry season maxima and minima of 36°C and 8°C. No recordings were made in 1974.

There was thus little difference between the maximum temperatures in each area, or between the minimum temperatures during the wet season. During the dry season in the south, however, minimum temperatures fell to a lower level than in the north so that the southern groups were exposed to wider fluctuations in daily temperature at that time.

c) *Daylight hours*

In both study areas, there was an approximately three-hour increase in the number of daylight hours during the wet season (Fig. 4).

2. Forest Structure

The estimated density of sampled plants in the two study areas was similar: 13,878 per hectare in the north and 13,370 per hectare in the south. There were striking differences between study areas in many other structural aspects of the vegetation:

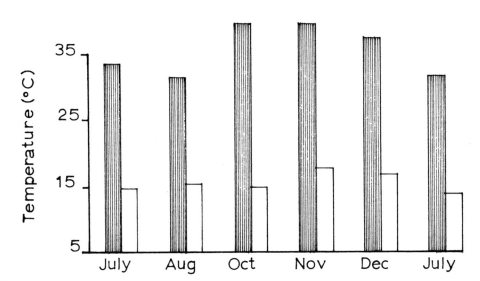

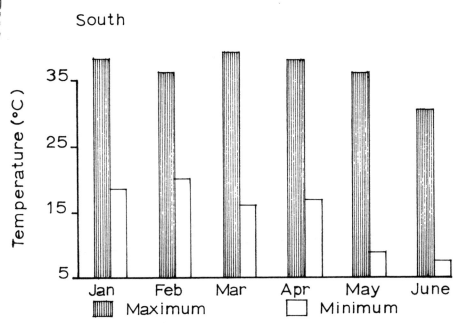

FIG. 3. Maximum and minimum temperatures recorded in each study area each month (1970-71).

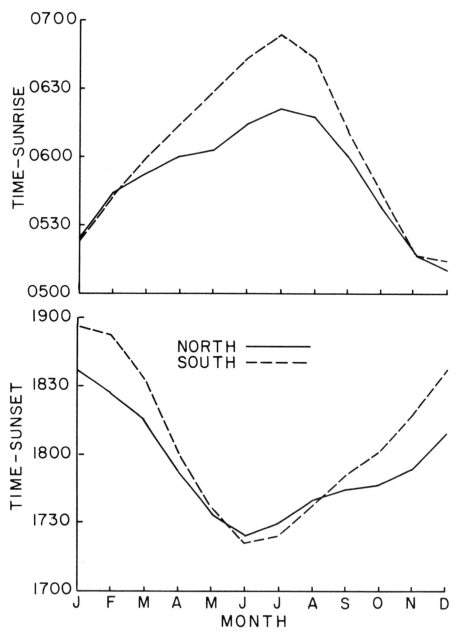

FIG. 4. Changes in the times of sunrise and sunset in each study area during a full
year.

a) *Ground layer*

The proliferation of tough woody lianas and shrubs in the north made the forest impenetrable to *P. verreauxi* and myself alike at ground level in some parts. In the south, the forest floor was generally covered by mosslike *Selaginella* species, and scattered with small woody shrubs. Rarely more than 1 m in height and usually growing singly or in small clumps, these shrubs provided no effective barrier to progressions on the ground; this sparseness may have been the result of extensive grazing. Further, there were nine low granite outcrops in the southern study area ranging from approximately 0.025 ha to 0.25 ha in size, which supported no vegetation (Plate VIII).

b) *Foliage*

Both the abundance and size of leaves were much greater in the north than in the south. The general reduction of foliage in the south may be an adaptive response to arid conditions; it is known that plants often adapt to chronic water shortage by extreme reduction or even disappearance of leaves (Kearney and Shantz 1911; Oppenheimer 1960; Schimper 1903; Treshow 1970).

c) *Tree height*

In the north, most trees were between 3 and 13 m in height, with emergents occasionally exceeding 21 m (Fig. 5). The small stature of trees was probably a result of impoverished sandy soil as well as of the age of the forest; except for the forest's low profile, the usual characteristics of a very young secondary succession (Richards 1966) were absent. In the south, most trees were between 3 and 7 m high, and none exceeded 21 m (Fig. 5). The difference between the distribution of trees among the various height categories in each study area shows that there were significantly more trees of greater stature in the north than in the south (Kolmogorov-Smirnov two-sample test, n_1=2,619, n_2=3,136, D=0.314, p<.001). It appeared that there was a critical size beyond which two of the most abundant *Alluaudia* species present in the south, *A. procera* and *A. ascendens*, became top-heavy, with the entire branching superstructure ultimately being torn off the trunk by its own weight. In the past, vegetation in this region has been referred to as *bush* (Cabanis et al. 1970). Here it is referred to as *forest*, partly to distinguish it from the low scrub-bush found over wide areas of southern Madagascar, and partly because *bush* seems an inappropriate term to describe vegetation containing trees up to 21 m in height.

d) *Tree spread*

In addition to being generally lower, trees in the southern study area provided less cover than in the north: summing the maximum

PLATE VIII. Adult female carrying infant on her back as she leaps bipedally across a
        rocky outcrop

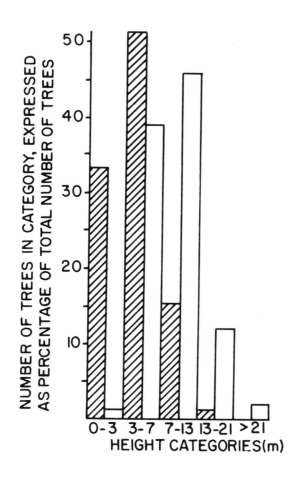

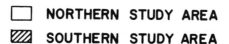 NORTHERN STUDY AREA

SOUTHERN STUDY AREA

FIG. 5. Distribution of sampled trees between height categories, within each study area.

linear spread of all trees in each plot of vegetation sampled, the mean summed spread of trees per sample in the south was 214 m. In the north it was 493 m.

e) *Lianas*

There were more lianas in the north than in the south. Lianas grew on an average of 70% of the trees sampled in each plot in the north, and on only 30% in the south.

3. Species Composition of Forests

A greater diversity of tree species was found in the north than in the south. A total of 194 species, from at least 35 families, was recorded in the north compared with only 99 species, from at least 23 families, in the south. Of these species, 177 in the north and 41 in the south fell within the plots of vegetation sampled. The remaining species were recorded when animals ate them. Tables 2.1 and 2.2 list species identified in the vegetation samples, together with their total frequency and the number of sample plots in which each occurred. For both areas sampled these figures undoubtedly underestimate the actual diversity of species; animals ate species that never fell within sample plots, and new species continued to be found in the plots themselves throughout the sampling of vegetation (Fig. 6). However, such species were probably widely scattered and rare in the forest.

TABLE 2.1

TREE SPECIES COLLECTED IN THE COURSE OF THE VEGETATIONAL ANALYSIS IN THE NORTHERN STUDY AREA, WITH SCIENTIFIC NAMES WHERE KNOWN, THE TOTAL NUMBER OF INDIVIDUALS OF EACH SPECIES COUNTED, AND THE NUMBER OF SAMPLE PLOTS IN WHICH THEY WERE FOUND (TOTAL NUMBER OF SAMPLE PLOTS = 24)

Family*	Genus & Species and/or given number	Total Number Found	Number of sample plots in which species was found
Euphorbiaceae	*Drypetes* sp. No. 18	323	24
Burseraceae	*Hymenodictyon* sp. No. 93	208	24
Rhizophoraceae	*Cassipourea* ? *phaeotricha* Tul.	116	13
Rhopalocarpaceae	*Rhopalocarpus similis* Hemsley	103	24

*No family is listed in cases where it was not given when the sample was identified by genus and/or species, and could not subsequently be traced by the author.

TABLE 2.1: Cont.

Family	Genus & Species and/or given number	Total Number Found	Number of sample plots in which species was found
Acanthaceae	No. 111	103	23
Rubiaceae	*Rothmannia decaryi*	85	19
Sapotaceae	*Capurodendron microlobum* (Baker) Aubreville	78	21
Leguminosae	*Baudouinia fluggeiformis* Baill	71	22
Anacardiaceae	*Protorhus deflexa* H. Perr	67	19
Myrtaceae	*Eugenia tropophylla* H. Perr	59	7
Tiliaceae	*Grewia* sp. No. 121	57	20
Euphorbiaceae	*Alchornea alnifolia*	56	17
Oleaceae	*Anacolosa pervilleana* Baill	56	21
Meliaceae	*Cedrelopsis* sp. No. 123	54	10
Burseraceae	*Commiphora pervilleana*	51	16
Guttiferae	*Rheedia arenicola* Jerm Perr	43	14
Ebenaceae	*Diospyros* sp. No. 126	41	5
Guttiferae	*Mammea* sp. No. 5	36	14
Anacardiaceae	*Poupartia* sp. No. 244	34	6
Apocynacae	*Hazunta* sp. No. 156	33	15
Rubiaceae	No. 276	32	11
Rhizophoraceae	*Cassipourea* sp. No. 155	31	10
Passifloraceae	*Adenia olaboensis* Clarerie	26	15
Sapindaceae	*Macphersonia gracilis* D. Hof.	25	10

TABLE 2.1: Cont.

Family	Genus & Species and/or given number	Total Number Found	Number of sample plots in which species was found
Sapotaceae	Mimusops sp. No. 320	24	8
Oleaceae	Linociera tropophylla H. Perr	23	13
Euphorbiacae	Euphorbia sp. No. 89	21	12
Convolvulaceae	Metaporana sp. aff. parvifolia (k. Afz.) Verdc	20	15
Annonaceae	Polyalthia Henricii Diels	19	8
Rutaceae	Vepris sp. No. 221	17	8
Canellaceae	Cinnamosma fragrans Baill.	17	10
Ochnaceae	No. 341	16	9
Celastraceae	Mystroxylon aethiopium	16	12
Sterculiaceae	Nesogordonia stylosa H. Perr	16	10
Capparidaccae	Boscia sp. No. 314	15	6
Verbenaceae	Premna longiacuminata	15	9
Papilionaceae	Dalbergia sp. No. 544	14	6
	No. 173	13	10
Papilionaceae	Mundulea sp. No. 64	13	6
?	Bathiorhamnus louveli H. Perr R. Cap. ssp. reticulatus R. Cap.	12	7
Euphorbiaceae	Phyllanthus sp. No. 113	12	4
Guttiferae	Psorospermun cerasifolium Baill	10	6
Boraginaceae	Vitex beraviensis	9	8

TABLE 2.1: Cont.

Family	Genus & Species and/or given number	Total Number Found	Number of sample plots in which species was found
?	*Mallaeastrum* sp. No. 240	9	6
Meliaceae	*Cedrelopsis* sp. No. 471	8	6
Rubiaceae	*Canthium* sp. No. 630	8	3
Myrsinaceae	*Ardisia didymopora* (Perr) R. Cap.	8	6
	No. 269	8	2
	No. 433	7	2
Myrtaceae	*Eugenia* sp. No. 603	7	3
	No. 358	7	4
Papilionaceae	*Indigofera* sp. No. 41	7	4
Boraginaceae	*Vitex* sp. No. 80	7	7
Ochnaceae	No. 97	7	7
Loganiaceae	*Strychnos madagascariensis*	7	6
Euphorbiaceae	*Sapium melanostictum*	7	5
Erythroxlyaceae	*Erythroxylon* sp. No. 530	7	5
	No. 183	6	9
Rubiaceae	*Enterospermum* sp. No. 81	6	9
	Holmskioldia microcalyx (J.F. Baker) Pieper	6	5
Loganiaceae	*Strychnos* sp. No. 142	6	4
	No. 336	6	5
	No. 260	6	2
	No. 339	6	1
	No. 492	6	3

TABLE 2.1: Cont.

Family	Genus & Species and/or given number	Total Number Found	Number of sample plots in which species was found
Tiliaceae	*Grewia* sp. No. 387	5	3
	No. 582	5	3
Sapindaceae	*Allophylus cobbe 'saligna'*	5	2
Tiliaceae	*Grewia picta* Baill.	5	2
Meliaceae	*Cedrelopsis grevei* Baill.	4	4
	No. 73	4	1
Mimosaceae	*Albizia arenicola* R. Viguier	4	3
	No. 622	4	2
	No. 229	4	2
Capparidaceae	*Boscia* sp. No. 403	4	2
	No. 511	4	1
Loganiaceae	*Strychnos descussata* Gilz	3	2
Verbenaceae	*Premna perplexans* Mold.	3	1
Euphorbiaceae	*Securinega seyrigii* Leandri	3	1
?	*Malhestrum aubingyense*	3	3
	No. 192	3	3
	No. 331	3	2
	No. 509	3	1
	No. 586	3	1
	No. 607	3	1
	No. 623	3	1
	No. 615	3	2
	No. 614	2	1

TABLE 2.1: Cont.

Family	Genus & Species and/or given number	Total Number Found	Number of sample plots in which species was found
	No. 618	2	1
	No. 619	2	1
	No. 344	2	1
	No. 600	2	2
Erythroxylaceae	*Erythroxylon* sp. No. 514	2	22
Pedaliaceae	*Uncarina* sp. No. 543	2	1
Capparidaceae	*Boscia* sp. No. 601	2	2
Annonaceae	No. 239	2	2
Rutaceae	*Vepris* sp. No. 263	2	1
Chrysobalanaceae	*Grangeria porosa* Boiv. ex Baill.	2	1
Sarcolaenaceae	*Perrierodendron boineuse* (Perr) Cav.	2	2
Guttifereae	*Psorospermum* sp. No. 130	2	1
Loganiaceae	*Strychnos* sp. No. 605	2	2
	No. 171	2	1
Capparidaceae	*Boscia* sp. No. 69	2	
Sapindaceae	*Allophylus* sp. No. 127	2	1
Compositae	*Brachylaena* sp. No. 165	2	2
Rutaceae	*Vepris* sp. No. 178	2	2
Rubiaceae	*Gardenia suavissima* Hom. ex Cavaco	2	2

One tree from each of the following species was noted:

Sapindaceae: *Allophylus cobbe "salicifolia"*; Rhamnaceae: *Ziziphus mauritania* Lamk, *Berchemia discolor*; Bignoniaceae: *Phyllarthron bernierianum*; Combretaceae: *Terminalia bovinii* Tul., *T. tropophylla*, Perrier; Annonaceae: *Xylopia sericolampra* (Dicks); Linaceae: *Hugonia Longipes* H. Perr.

TABLE 2.1: Cont.

One tree from a species within each of the following families and/or genera was noted:

Acanthaceae; Rhizophoraceae, *Cassipourea*; Ebenaceae, *Diospyros*; Meliaceae; Euphorbiaceae, *Drypetes, Croton*; Erythroxylaceae, *Erythroxylon*; Humberticoturraceae; Myrsinaceae, *Ardisia*; Burseraceae, *Commiphora*; Rubiaceae, *Canthium*; Meliaceae, *Turraea*; Papilionaceae, *Indigofera*.

In addition, 47 totally unidentified samples were collected which came from what appeared to be 47 different species.

TABLE 2.2

TREE SPECIES COLLECTED IN THE COURSE OF THE VEGETATIONAL ANALYSIS IN THE SOUTHERN STUDY AREA, WITH SCIENTIFIC NAMES WHERE KNOWN, THE TOTAL NUMBER OF INDIVIDUALS OF EACH SPECIES COUNTED, AND THE NUMBER OF SAMPLE PLOTS IN WHICH THEY WERE FOUND (TOTAL NUMBER OF SAMPLE PLOTS = 30)

Family	Genus & Species and/or given number	Total Number Found	Number of sample plots in which species was found
Didieriaceae	*Alluaudia procera*	916	30
Burseraceae	*Commiphora humbertii* H. Perr	607	29
Didieriaceae	*Alluaudia ascendens*	251	24
Meliaceae	*Cedrelopsis grevei* Baill.	211	23
Combretaceae	*Terminalia* sp. No. 048	132	27
Buseraceae	*Commiphora orbicularis* Engler	117	22
Euphorbiaceae	*Euphorbia plagiantha*	115	10
Burseraceae	*Commiphora* sp. No. 076	72	21
Ebenaceae	*Diospyros humbertiana* H. Perr	66	13
Burseraceae	*Commiphora simplicifolia* H. Perr	66	23
Anacardiaceae	*Operculicarya decaryi* H. Perr	56	23
Euphorbiaceae	*Securinega* sp. No. 016	50	13

TABLE 2.2: Cont.

Family	Genus & Species and/or given number	Total Number Found	Number of sample plots in which species was found
Tiliaceae	*Grewia* sp. No. 089	35	16
Hernandiaceae	*Gynocarpus americanus* Jacq.	34	11
	No. 015	27	17
Euphorbiaceae	No. 080	27	11
Euphorbiaceae	No. 098	25	10
Didieriaceae	*Alluaudia humbertii*	21	6
Mimosaceae	*Mimosa* sp. No. 033	21	9
Cesalpiniaceae	*Tetrapterocarpum geayi* H. Humb.	20	11
Burseraceae	*Hymenodictyon* sp. No. 975	19	15
Combretaceae	*Terminalia* sp. No. 0109	19	5
Euphorbiaceae	*Croton* sp. No. 062	18	8
Rubiaceae	*Enterospermum* sp. No. 085	18	13
Apocynaceae	*Hazunta modesta* (Bak.) M. Pichon	17	9
Combretaceae	*Terminalia* sp. No. 09	16	8
Tiliaceae	*Grewia* sp. No. 059	14	7
Tiliaceae	*Grewia* sp. No. 0106	14	5
Bignoniaceae	*Stereospermum nematocarpum*	11	6
Leguminaceae	No. 051	11	2
Meliaceae	*Neobegnea mahafalensis* J. Leroy	10	8
Bignoniaceae	*Rhigozum madagascariense* Drake	10	4

TABLE 2.2: Cont.

Family	Genus & Species and/or given number	Total Number Found	Number of sample plots in which species was found
Ebenaceae	*Diospyros* sp. No. 013	9	6
Euphorbiaceae	*Euphorbia stenoclada*	9	4
Salvadoraceae	*Salvadora angustifolia* Turill.	8	5
	No. 099	8	7
Burseraceae	*Commiphora pervilleana* Perr	6	5
Cesalpiniaceae	*Baudouinia fluggeiformis*	5	3
Tiliaceae	*Grewia* sp. No. 0126	5	2
	No. 026	5	1
Burseraceae	*Hymenodictyon* sp. No. 093	4	4
Burseraceae	*Commiphora brevicalyx* H. Perr	3	3
	No. 0100	3	1
Papilionaceae	*Indigofera* sp. No. 0111	3	1
	No. 0121	3	1
	No. 095	3	1
Mimosaceae	*Albizia* sp. No. 084	2	1
Bombacaceae	*Adansonia fony*	2	2
Malvaceae	*Hibiscus* sp. No. 0116	2	2
Apocynaceae	*Pachypodium lamerei* Drake	2	2
Mimosaceae	*Entada abyssinica* Stendel	2	2

TABLE 2.2: Cont.

Family	Genus & Species and/or given number	Total Number Found	Number of sample plots in which species was found
Rubiaceae	*Enterospermum* sp. No. 0120	2	2
	No. 0130	2	1
	No. 045	2	2
Capparidaceae	*Boscia longifolia* Hadj. Monst.	2	2
Apocynaceae	*Rauwolfia confertiflora* M. Pichon	2	2
Urticaceae	*Pouzolzia gaudichaudii* J. Leandri	2	2

One tree from each of the following species was noted:

Verbenaceae: *Vitex microphylla* Moldenke; *Rothmannia decaryi*; Anacardiaceae, *Rhus thouarsii* H. Perr.

One tree from a species within each of the following families and/or genera was noted:

Euphorbiaceae, *Croton* (three different samples); Ebenaceae, *Diospyros;* Mimosaceae, *Acacia;* Bignoniaceae, *Stereospermum;*

In addition, five totally unidentified samples were collected which came from what appeared to be five different species.

In both forests, individual trees from most species were found singly in one or a few sample plots, and only a small number of species were present in all or many of them. In Figure 7, species are grouped according to the number of sample plots in which they occurred. Thus, for example, in the north 62% of species occurred in only 10% of the sample plots (although not necessarily the same 10%; these figures refer only to frequency of occurrence and not to distribution). There was a close correlation between the number of sample plots in which a species occurred and the total number of individuals of that species counted.

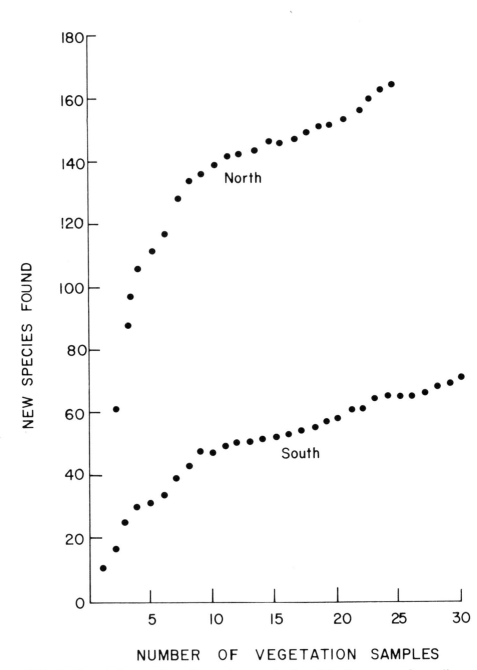

FIG. 6. Cumulative number of new tree species found in the course of sampling
    vegetation in each study area.

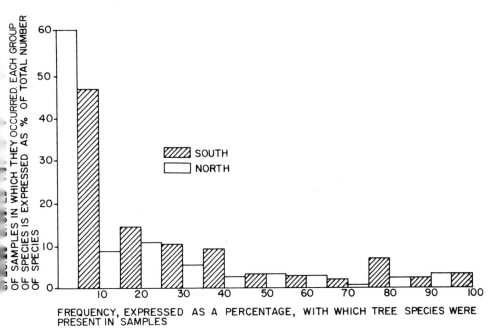

FIG. 7. Tree species grouped according to the number of sample plots in which they occurred, in each study area.

Concerning the distribution of tree species, therefore, it can only be said that a few species were abundant and widely, though not necessarily uniformly, distributed through the forest in each study area. Most species occurred rarely, unevenly scattered among the few dominant species.

4. Phenology

The phenology of ten species was described each month of observation in either study area (Figs. 8 and 9). Ten individuals were sampled from each species. These data were used to estimate changes in the abundance of leaves, flowers and fruit, and so on, both among species within each forest and among individuals of each species. In this way, a systematic comparison could be made of the "seasonality" of each study area, or the amount of seasonal variation in the overall presence of different food parts. The general phenology of both study areas showed similar changes between seasons (Fig.10); a peak in the frequency of immature

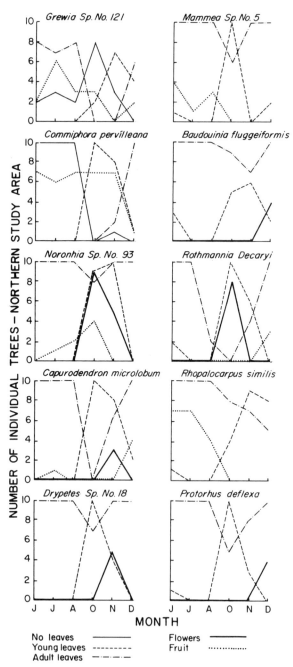

FIG. 8. Phenology of trees sampled from ten species in the northern study area during the study period.

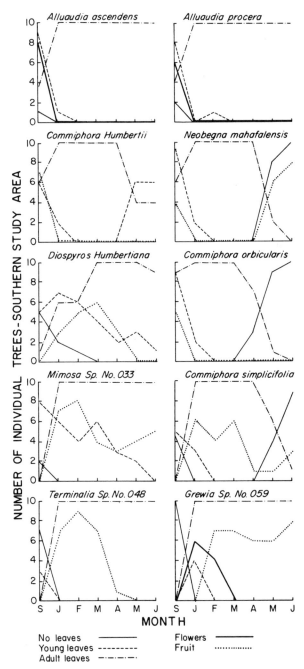

FIG. 9. Phenology of trees sampled from ten species in the southern study area during the study period.

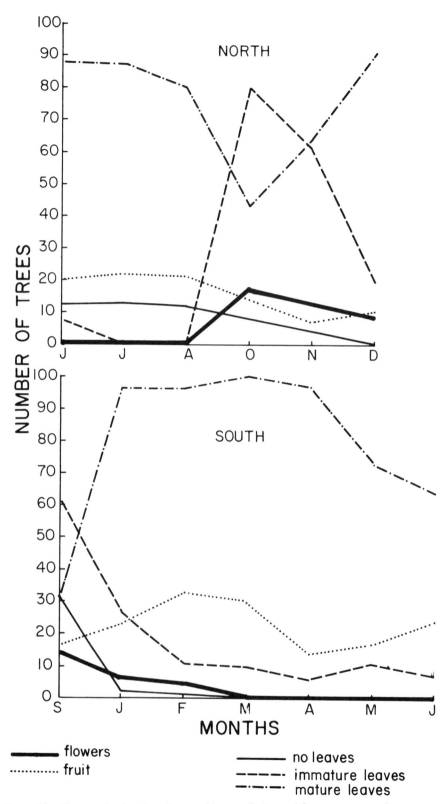

FIG. 10. Changes in the abundance of leaves, fruit, and flowers among the ten tree species sampled in each study area.

PLATE IX. Seasonal change in the northern forest

leaves being recorded present coincided with a decrease in the presence of mature leaves. In the north this occurred at the onset of the wet season in October; in the south it occurred after light rainfall in September 1970, although the real wet season did not begin until January. In the north, flowers were most abundant in October, and as the frequency of occurrence of flowers declined in December, the number of fruiting trees began to increase. In the south, similarly, fruit was most commonly found towards the end of the rains in February and March. The number of trees without leaves was at its height in north and south at the end of the dry season.

The south was an area of extreme seasonality of climate, with long periods of drought. The north had a more moderate climate, yet seasonal changes in vegetation were no less marked than in the south. At the beginning of the dry season, many trees in the north lost their leaves and production of deciduous leaves stopped, although evergreen species retained their full foliage (Plate IX). By contrast, although the aridity of the southern dry season was much greater, some trees of many species continued to produce new leaves throughout the study period (Fig. 10). This pattern differed from that of true evergreens in that not all individuals of these species produced dry-season foliage (Fig. 9) and none maintained it throughout the dry season. It is probably the evolution of water-conserving adaptations by xerophytic species in the south that enables them to continue to produce leaves well into the dry season instead of lying dormant for the nine dry months of the year.

Notes to Chapter 2

1. Jolly (1966) states that "the 21-month animals are indistinguishable from adults," but the short, highly synchronized breeding and birth season each year and the three distinguishable (by size) grades of immature animals suggest that maturation must take about three years (cf. Altmann et al., in press).

2. These calls are described in detail by Jolly (1966), and Petter (1962a) shows sonograms of the "sifaka" and the "roaring bark." No systematic analysis has been undertaken of data collected on the frequency of these calls, or of tape recordings made in the course of the study, but the context and postulated function of calls are discussed in the text where appropriate.

3
Group Size, Composition, and Changes through Time

1. Areas Surveyed

Limited surveys were conducted in the south and southwest of the island, at Evasy and Ejeda (Fig. 1). In the study areas, the composition of neighboring groups was established when they encountered the study groups. The southern study groups were recensused in 1974. The most thorough censuses were carried out in 1974 at Antserananomby, and in 1971 and 1974 at Berenty. This reserve was selected for repeated censuses because information was thereby provided not only on group size and composition, but also on changes in these factors through time; these were the fourth and fifth censuses to be made on this population in eight years (see Jolly 1966, 1972).

The forest at Evasy was semi-arid, similar in structure and composition to the southern study area. At Ejeda, on the River Linta, Berenty, on the Mandrary, and Antserananomby, on the River Bengily (a tributary of the Mangoky), the habitats surveyed were gallery forests, dominated by kily trees, *Tamarindus indica*.

2. Group Size and Composition

Group size ranged from 3 to 12, with mean group size ranging from 5.0 (in the northern study area) to 7.8 (at Antserananomby). The results

67

of these area censuses are summarized in Table 3.3. There was no statistically significant difference between the size of groups in the forests at Ampijoroa, Hazafotsy, Berenty, Evasy, and Ejeda. The groups at Antserananomby were significantly larger than those at Berenty, where mean group size was next highest (Mann Whitney U test, U=83, p <.001). (Comparisons of group size and the calculation of mean group size were made excluding infants, in order to compensate for seasonal differences in the timing of surveys.)

The overall adult sex ratio was greater than one (i.e., females outnumbered males) in all forests except Antserananomby, but the range in this ratio between groups in each forest was wide; at Ampijoroa, for example, one group contained two adult males and five adult females, while another contained three males and only one adult female.

TABLE 3.3

COUNTS, PARTIAL ANALYSIS OF THE AGE AND SEX COMPOSITON AND MEAN GROUP SIZE IN STUDY AREAS, AND AT BERENTY, EVASY, EJEDA, AND ANTSERANANOMBY

Locality & Date	Adult Male	Adult Female	Sub-adult	+	Juv.**	Inf.	Total
Ampijoroa (northern	*2	5					7
study	*1	1	1(♂)		1(♂)	1	5
area)	3	1					4
July '70	3	2		1		1	7
	1	2		1		1	5
	2	3					5
	3	5		1		1	10
	15	19		5		4	43
							6
		Adult					5
		Composition				1	4
		Unknown					4
						1	4
Total number in 12 groups							66

Ampijoroa: Mean group size 5.0
Adult sex ratio   1:1.3

* Study groups
**Subadults and juveniles were distinguished only in the study groups.

TABLE 3.3: Cont.

Locality & Date	Adult Male	Adult Female	Sub-adult	+	Juv.**	Inf.	?	Total
Hazafotsy	*2	2	1(♂)		1(♂)	2		8
(southern	*1	2	1(♂)			2		6
study	2	1				1		4
area)	1	2		1		1		5
Sept. '70	2	4		1		1		8
	8	11		5		7		31
								8
		Adult						6
		Composition						5
		Unknown						5
								3
Total number in 10 groups								58

Hazafotsy: Mean group size 5.1
Adult sex ratio 1:1.4

Locality & Date								Total
Evasy May '70								10
		Adult						7
		Composition						7
		Unknown						6
								5
								4
								4
								4
Total number in 8 groups								47

Evasy: Mean group size 5.9

Locality & Date								Total
Ejeda May '70		Adult						7
		Composition						7
		Unknown						
Total number in 2 groups								14

* Study groups
**Subadults and juveniles were distinguished only in the study groups.

TABLE 3.3: Cont.

Locality & Date	Adult Male	Adult Female	Subadults & Juveniles	Inf.	?	Total
Berenty	3	2	1	2		8
Sept. '71		1	1	1	1	7
	3	1				3
	3	3			S-a:1	7
	1	5		1	A:1	8
	3	2				5
	2	2	1	2		7
	2	4	1	2		9
	3	3	1	2		9
	2	1	1		A:1	5
	(*2)					(2)
Totals for 10 groups	24	24	6	10	4	68

Berenty: Mean group size 5.8
Adult sex ratio 1:1

*These two males moved together in the area occupied by the ten groups listed.

Locality & Date	Adult ♂	Adult ♀	Subadult & Juveniles	Infants	Total
Berenty	2	3	1♂		6
May &	1	5	1♀		7
June '74	2	2			4
	3	1	1♂		5
	2	2	1♂		5
	2	1	1♂		4
	4	2	1♀		7
	2	1	1♀		4
	2	4	1♂		7
	2	2			4
	1	2			3
	2	2			4
	2	3	1♂		6
	3	3	1♀		7
	1	2			3
	2	2	1		5
	1	3			4
	2				2
	36	40	7♂ + 4♀		87

Berenty: Mean group size 5.0
Adult sex ratio   1:1.1

TABLE 3.3: Cont.

Locality & Date	Adult Male	Adult Female	Subadults & Juveniles	Infants	Total
Antseran-	5	5	1♂	1	12
anomby	5	4	1♀	1	11
July 1974	4	5	1♀	1	11
	4	4	1♀	1	10
	2	1		1	4
	4	1		1	6
	2	2		1	5
	4	4	1♂	1	10
					7
		Adult composition unknown			10
	30	26	2♂+3♀	8	86

Antserananomby: Mean group size 7.8
Adult sex ratio 1:0.9

Hazafotsy	3	1	1(juvenile)		5
June 1974	1	3	1(juvenile)		5
	4	4	1		10

Hazafotsy: Mean group size 5

3. Births and Reproductive Cycles

The changes in group composition discussed in this and Section 4 are summarized in Table 3.4.

TABLE 3.4

INITIAL COMPOSITION OF THE FOUR STUDY GROUPS AND SUBSEQUENT CHANGES

Group	July 1970	Oct. 1970	Dec. 1970	July 1971	June 1974
I	5 A♀♀	5 A♀♀	4 A♀♀	4 A♀♀	—
	A♂ GOP	A♂ STR	A♂ BE	A♂ BE	
	A♂ N			2 Infs.	
II	A♀	A♀		A♀	
	A♂	A♂		A♂	
			No Change		
	SA ♂	SA♂		SA ♂	
	J	J		J	
	Inf.			Inf.	

TABLE 3.4: Cont.

	Sept. 1970	March 1971*	June 1971	Sept. 1971	June 1974
III	2 A♀♀	2 A♀♀		2A♀♀	1 A♀
	2 A♂♂	2 A♂♂		2A♂♂	3 A♂♂
			No Change		
	SA ♂	SA ♂		SA ♂	
	J	J		J	J
	2 Infs.			2 Infs.	
IV	2A♀♀	2 A♀♀		2A♀♀	3A♀♀
	A♂ R	A♂ INT		A♂ INT	1A♂
			No Change		
	SA ♂	SA ♂		SA ♂	J
	2 Infs.			Inf.	

*A♂ P's brief attachment from Group III and association with Group IV is not recorded here. Where group members are not individually named, it can be assumed that their identity remained the same.

a) *Births* (see also, Richard 1976)

i) Northern study area

One infant was born in Group II in 1970, but no infants were born to any of the five adult females in Group I that year. Five other infants, each in a separate group, were known of in the area. These births occurred during June and July. When observations were taken up in October after a six-week interval, the Group II infant and all but one of the other infants had disappeared. The infant mortality rate for 1970-71 was therefore approximately 84% (n=6).

In 1971 the Group II female produced another infant, and two females in Group I produced infants. Since observations ended in the north in July 1971, the survival rate of these 1971 infants is not known.

ii) Southern study area

In September 1970 all the adult females in Groups III and IV were found with infants. These infants had probably been born toward the beginning of August; this was inferred from their size and relative independence. When observations were resumed in January 1971, only one infant remained, belonging to adult female FI in Group IV. This infant finally disappeared too during a cyclone in mid-February. Infant mortality in the groups sampled in 1970/71 was thus 100% (n=4).

In August 1971 infants were born to the two adult females in Group III but only to adult female FI in Group IV. Here again, it is not known whether these infants survived.

The 1974 birth season had not yet occurred when observations were ended in the southern study area on June 19.

b) *Reproductive cycles*

Petter-Rousseaux (1962) has shown that the Cheirogaleineae are seasonally polyoestrous in the laboratory. Evans and Goy (1968) have shown that this is also true for *Lemur catta*. Jolly (1966) produced evidence for *L. catta* in the wild having a "pseudoestrous period" approximately one month earlier than the true breeding season; the vulval area of four or five of nine females went through a pink phase 3 to 4 weeks before the week of mating, faded, and then flushed again just prior to mating. Two of these females were seen mating in the second period of flushing.

Data on the timing of births indicate a single full oestrus and close synchrony among females within and between groups. *P. verreauxi* has a gestation period of about 130 days (Petter-Rousseaux 1962). The length of the oestrous cycle is not known, but it is probably about the same as *L. catta*'s in captivity: 39.3 days, with a range of 33-45 days (Evans and Goy 1968). If this is so, and if females were seasonally polyoestrous with a closely synchronized cycle, two birth peaks would be expected, the second about five weeks after the first, when females fertilized during the second oestrous period gave bith. However, in the northern study area, births of *P. v. coquereli* were scattered over a maximum of 21 days in 1970 and 1971. Jolly (1966) reported a ten-day birth season for *P. v. verreauxi* at Berenty, in the south of the island. In neither case was the distribution of births through time sufficient to indicate a polyoestrous breeding system.

Data from the mating season (see also chapter 5) are consistent with those collected on births. Observations suggest that *P. verreauxi* does not have more than one full oestrus each year, but that there is a period analogous to the "pseudoestrous period" in *L. catta*: flushing of the vulva of one of the females in Group IV and associated activities in late January probably represented a partially suppressed oestrus 37 days before full oestrus. This did not occur in Group III, despite the presence of two adult females, so presumably the suppression was total in this group. Since both females in Group III gave birth that year, they must have come into full oestrus once, probably at the same time, either between February 13 and 20 (when observations were interrupted by flooding) or after March 15 (when they were temporarily abandoned because of political unrest in the area).

The similarity in cycling between *L. catta* and *P. verreauxi* is substantiated by other aspects of their mating systems; the receptivity of Group IV females FNI and FI lasted a maximum of 42 and 36 hours respectively, as compared with a maximum of 36 hours for *L. catta* in the wild (Jolly 1966) and 10 to 24 hours in captivity (Evans and Goy 1968). The incidence of marking behavior by females in this study was not affected by the breeding season, and a similar stability is reported in *L. catta* females in captivity (Evans and Goy 1968). Finally, Evans and Goy report: "Both long and short term fluctuations in gonadal activity were associated with changes in the frequency of expression of several non-sexual patterns . . . , " although they do not differentiate clearly between *sexual* and *nonsexual*. In *P. verreauxi*, changes in the frequency of several activity patterns occurred during the mating season and, as in *L. catta*, only two new behaviors appeared: the act of copulation itself, and fierce fighting between adult males. It is probable that these frequency changes were, similarly, associated with changes in gonadal activity.

P. verreauxi mated at the end of the wet season, females gestated until the middle of the dry season, and infants were weaned by the beginning of the following wet season. In 1965 Lancaster and Lee surveyed research done on ultimate and proximate factors determining the timing of primate reproductive cycles, their primary conclusion being that more research was needed. The effect of the photo-period on oestrus in *Microcebus* has been demonstrated by Petter-Rousseaux (1969). The breeding season of *L. catta* shifts by six months when animals are brought to the northern hemisphere (Martin pers. comm.), but controlled experiments have yet to be done on either *L. catta* or *P. verreauxi* to establish the precise influence of day length and temperature on their sexual behavior. Since *L. catta* is polyoestrous in captivity and not in the wild, it is likely that, in this species at least, proximate ecological factors are important in the wild and it is their absence in captivity that permits the appearance of the full polyoestrous system.

4. Disappearances and Possible Causes of Death

Most infants born during the study disappeared within a few months of their birth and are presumed to have died. One adult female is known to have died in Group I, and the fur of a second animal was found in the range of this group. Shortly before this find an adult male disappeared from the group, and it is probable that these were his remains. Two more adult males disappeared from this group during the study, but there is no evidence to suggest that they died. The composition of Group II remained constant throughout the study. The composition of Group III did not change during the 1970/71 study, but when the group was recensused in 1974 one adult female and one adult male had disappeared. The adult male in Group IV was driven out during the mating season by an "invading" male; he was severely wounded during the conflict and never seen again (see Richard 1974). In 1974 there was only one adult male in the group (this was certainly not A♂ INT of 1971, but resembled in coloring and facial features the subadult male SA♂Q, present in the group in 1971. Although the identification is only tentative, it is probable that it was A♂ INT who had disappeared from the group).

Some of these disappearances may have been due to the death of the animal in question. Factors that may be pertinent to mortality rates in *P. verreauxi* are considered below.

a) *The significance of "alarm" calls*

The study groups' vocal repertoire included one strident call, which appeared to function as an alarm call. This "roaring bark" was usually initiated by one animal and then taken up almost at once by the whole group, and occasionally by neighboring groups. It was generally given when an aerial predator flew overhead, and was often accompanied by a locomotor response; as the call was given, animals moved rapidly from the periphery of the tree to the trunk. They then backed down toward the ground, looking up as they did so. A second call, the *sifaka*, was generally given in response to the presence of potential predators on the ground: it was always directed at me by unhabituated groups. On the two occasions when my assistant was attacked by unhabituated groups, however, the roaring bark was given. Further, on one occasion when animals found a snake strangling an adult *Microcebus murinus*, the animals gave a roaring bark. The specificity of these calls is thus unclear, although Jolly (1966) postulates that the roaring bark is directed at raptors, whereas the sifaka is given in response to ground predators.

The presence of these alarm calls cannot be taken as conclusive evidence of predator pressure, despite Cartmill's (1972) comment: ". . . unless these calls (roaring alarm barks) are another Eocene survival, these Malagasy VCLs must be subject to predation." Both chimpanzees and gorillas have alarm calls, although predation pressures are probably minimal (Van Lawick Goodall 1965; Schaller 1963), and there is evidence that alarm calls in *P. verreauxi* may function to draw attention to sudden, unexpected stimuli as well as to predators *per se* (cf. the howls of howler monkeys, *Alouatta palliata*, Baldwin and Baldwin, in press); for example, the northern study groups always gave a roaring bark when trucks passed on the road 3 kms from the study area.

In sum, the presence of alarm calls in *P. verreauxi* cannot be used alone as grounds for suggesting that the deaths and disappearances that occurred during the study were due to predation.

b) *Possible nonhuman predators*

From observations in this study, it is inferred that animals, particularly infants, may have three nonhuman predators: raptors, constrictor snakes, and the Viverrid *Cryptoprocta ferox*, known in Madagascar as the "Fossa".

About one month after birth, infants grew more and more adventurous and their mothers less and less attentive, with the result that the two might become separated by as much as 4 to 5 m. The infants were still ungainly and slow-moving at this time, and appeared to constitute an ill-camouflaged sitting target for raptors. The vocal and locomotor response of all animals to the presence of these birds lends support to this possibility. There were two species of large raptor in both study areas: *Polyboroides typus* and *Gymnogenys radiata*.

A snake, *Ithycyphys miniatus*, was noticed by Group I in the northern study area while it was strangling an adult *Microcebus murinus*. After giving the roaring bark alarm, the group sat down about 7 m from the snake and watched silently for 25 minutes while the snake swallowed its catch. The group then moved away and resumed feeding. It is possible that *P. verreauxi* infants, at about one month, could also be vulnerable to such snakes.

A third potential predator on infants, and possibly even on adults, particularly the old or diseased, is *Cryptoprocta ferox*. Nests of Fossa young have been found in Baobab trees in the southwest, with *P. verreauxi* bones strewn at their foot (Albignac pers. comm.). These could have been scavenged, but eye-witness reports given by two villagers spoke of prolonged fights between adult *P. verreauxi* and the Fossa.

c) *Relations with man*

Undoubtedly the most important single predator of adult *P. verreauxi* is now man. Although protected by law, these animals are hunted for sport and for food in many areas (Richard and Sussman 1975).

Groups made two attacks on my assistant (cf. Rand 1936); on both occasions he had stalked an unhabituated group to within a few meters. It is plausible that since he, as "predator," had approached to within the groups' flight distance, their response was to attack rather than to flee. There was, I think, no doubt that in both instances the groups pursued and attempted to attack, rather than simply mobbing my assistant before ultimately being repelled with stones.

d) *Disease*

In October 1970 a female aged 3-4 years from Group I was found sitting curled up on the ground. She offered no resistance to being picked up by me, and died two days later. Subsequent bacteriological analysis did not reveal the cause of death, although numerous nonpathogenic protozoans were found in the red blood cells in the kidney (for a further discussion see Uilenberg et al., in press).

The disappearance and presumed death of the infant in Group IV during a cyclone may have been related to disease. Temperatures remained low continuously for about a week at that time, and it rained incessantly. It is possible that the infant died from exposure to cold directly, or that the cold indirectly killed it by reducing its resistance to viral infection.

5. Movement of Animals between Groups

No females or immature animals were seen leaving or entering the study groups, but a number of changes occurred in the male composition of the groups, both in and out of the mating season.

During the mating season in the south, adult male R was ousted from Group IV (see chapter 5). His successor was in turn ousted by another intruding male, who remained in Group IV until the study was terminated the following September. In June 1970 two adult males were present in Group I. By October both had disappeared from the group and a new male was present, adult male STR. At the beginning of December another male, BE, began moving on the periphery of Group I; as with adult male STR, it was not known whether he had come from another group or was a solitary, wandering male. Members of Group I made no attempt to chase him away, nor did he try to drive off STR. When observations were subsequently resumed after a week's interval, STR had disappeared from Group I and BE was well integrated into the group. There was no evidence in the form of recent wounds to suggest that he had had to fight his way into the group. Adult male INT's disappearance from Group IV between 1971 and 1974 could be interpreted in this way — as a change of group rather than as a death.

6. The Population of *P. verreauxi* at Berenty

a) *Stability of groups*

The size, composition, and range of six known groups changed little

TABLE 3.5

COMPOSITION OF KNOWN GROUPS AT BERENTY, TAKEN FROM CENSUSES MADE BY JOLLY IN SEPTEMBER 1970 AND BY STRUHSAKER AND RICHARD IN SEPTEMBER 1971

Year	Group's Name	Adult Male	Adult Female	Juv.	Inf.	Total
'70	WP	2	4		2	8
'71		3	3			6
'70	DE	2	2		1	5
'71		2	2	1	2	7
'70	BN	3	3		2	8
'71		3	2	1	2	8
'70	PT	3	2		2	7
'71		3	3	1	2	9
'70	ST	2	1			3
'71		2	1			3
'70	SE	2	1			3
'71		2				2
TOTAL, 1970		14	13	—	7	34
TOTAL, 1971		15	11	3	6	35

between 1970 and 1971. Table 3.5 compares some of the results of Jolly's census in 1970 (Jolly 1972) with the census carried out with the assistance of T. T. Struhsaker in 1971. Groups from 1970 were reidentified in 1971 by matching up the geographical location of their home-ranges in the two years, and by identifying certain "marker" animals, described in detail by Jolly (pers. comm.). Only the six groups whose 1970/71 identity was confirmed by these two criteria are included in Table 3.5. The similarity of the 1970/71 results agrees with Jolly (1972), who returned to Berenty in 1970 after a six-year absence and found that the number, composition, and spacing of troops, as well as spacing behavior, seemed unchanged .

While the evidence from Berenty suggests that there was considerable stability in group size and age/sex composition, it does not in most cases demonstrate whether the actual identity of animals constituting groups remained the same. Jolly (1972) reported: "Four animals [two adult males and two adult females], out of a total of 36 known in 1963/1964 could be recognized in 1970, and were in the same home-ranges, including a mother who had infants in 1963, 1964, and

1970 . . . , " but the direct and indirect evidence for adult male mobility found in this study suggests that at least some adult males at Berenty may move between groups over a period of years.

b) *Population changes, 1963-1974*

The aim of the 1974 census was not to reidentify particular groups but to gather comparative material on a wider sample of groups. Comparing results of the censuses made in 1963, 1964, 1970 (Jolly 1966, 1972), 1971 (by Richard and Struhsaker), and 1974, the similarity in most measures suggests a high degree of stability in the population (Table 3.6). The only consistent change to have been taking place since 1963 is the gradual equilibration of the sex ratio. The overall estimated density was lower in 1974 than in previous years. This was probably because the 1974 census included areas of the forest further from the river where the vegetation was degraded; the density of animals in the 10 ha adjoining the river (where previous censuses were conducted) was comparable to that for previous years. Budnitz and Dainis (1975) have noted a similar effect for *L. catta* and cautioned against extrapolating overall population density in the Berenty Reserve on the basis of samples taken in only one limited area of the forest.

The mean number of offspring surviving one to two years per female per annum was iniformly low: 0.1 - 0.3. This accords well with the high infant mortality rates found in the two study areas, although it should nonetheless be considered a minimum estimate; in some instances, two- to three-year-old subadults may have been recorded as adult females. Indeed, in 1970 Jolly did not discriminate between juveniles and adult females.

c) *The changing sex ratio*

The excess of males over females at Berenty in 1963 and the subsequent equilibration of the sex ratio warrant special consideration. In 1963 Jolly counted 23 males and 15 females in 10 groups; this sex ratio of 0.65 has since modified to 0.71 (1964), 0.96 (1970), 0.91 (1971), and 1.02 (1974). In 1972, Jolly noted: "In 1970 as in 1963-1964 there were more adult males than adult females, a peculiarity which remains unexplained." In 1966 she was less cautious and proposed that the discrepancy in the sex ratio might be a consequence of males living longer than females. This theory was based on the presence of two "old" males who were balding, and had deeply sunken faces in the 1963/64 study. No females were seen in this condition. Basilewsky, in a discussion of captive breeding of lemurs, noted (1965): "The sex ratio in the group always shows a significant majority of males. This can be established without any doubt, particularly easily in *L. macaco* with its

TABLE 3.6

INFORMATION EXTRACTED FROM DATA COLLECTED DURING CENSUSES MADE IN 1963, 1964 (JOLLY 1966, 1972), 1970, 1971 (BY STRUHSAKER AND RICHARD), AND 1974

Parameter	Year				
	1963	1964	1970	1971	1974
Area censused (in hectares)	10	10	10	10	24
No. of groups counted	9	9 plus 1A♂	9 plus 1A♂	9 plus 2A♂♂	17 plus 2A♂♂
Mean group size	4.67	5.0	5.0	5.6	5.0
Variation in group size	2-6	2-8	3-6	3-7	3-7
Mean sex ratio	1:0.65	1:0.71	1:0.96	1:0.91	1:1.02
Range of group sex ratios	1:0.3 to 1:1	1:0.6 to 1:1	1:0.6 to 1:2	1:0.3 to 1:5	1:0.25 to 1:6
No. of infants counted	7	Prebirth season	10	10	Prebirth season
No. of juveniles counted	8	5	?	6	11 (including subadults)
Mean reproductive success per female per annum (i.e., mean number of surviving offspring per female per annum) $\frac{(\text{juveniles \& subadults})}{\text{total no. females x 2}}$	0.3	0.1	?	0.1	0.3
No. of animals counted (excluding infants and including solitary males)	42	46	46	53	87
Estimated density (No. of animals per hectare)	4.2	4.6	4.6	5.3	3.62
Estimated density in 10 ha adjoining river (No. of animals per hectare)	4.2	4.6	4.6	5.3	4.8

marked sexual dimorphism, but it is also true for *L. catta* and most other species." He did not specify whether this referred to the sex ratio at birth nor did he offer any explanation for this observation.

The results of this study confirm Petter's (1962a) findings of an approximately 1:1 sex ratio in *P. verreauxi*. They suggest that general theories of greater male longevity, differential infant mortality, or an unequal sex ratio at birth are superfluous and that the imbalance at Berenty in 1963 was a temporary and atypical phenomenon. First, the addition of census data for 1971 and 1974 confirmed and extended the trend in the sex ratio that appeared to be taking place between 1963 and 1970; after a gradual increase during the previous eleven years, it finally exceeded one for the first time, in 1974. Second, the sex ratio was more than one in all other populations surveyed during this study, except for that at Antserananomby, where it was 0.91. In a study of the population dynamics of *Papio anubis*, Rowell (1969) documented and emphasized the magnitude of naturally occurring fluctuations in the sex ratio of undisturbed groups from year to year. The sex ratio at Antserananomby can well be viewed as a low point in a cycle of normal fluctuations in that population.

The causes of the extremely imbalanced ratio at Berenty in 1963/64 remain obscure, although there is one possibility worth mentioning. Prior to 1963 extensive felling was carried out in the Mandrary River valley to clear land for sisal production. The reserve is itself surrounded by a "sea" of sisal. Evidence from this study suggests that males are more mobile than females and show less attachment to their range. Thus, it is possible that when the surrounding forest was felled, males took refuge in the reserve that was left intact whereas females remained within their ranges and were, presumably, killed. Over the years the males from this initial influx have gradually died and the sex ratio has been once more returned to equilibrium.

4
Maintenance Behaviors

1. The Home-Range

a) *Definition and mapping*

The area over which all four study groups ranged was divided into 50 m-sided quadrats. Jewell (1966) restated Burt's (1943) definition of home-range as follows: " . . . home-range is the area over which an animal normally travels in pursuit of its routine activities." In an attempt to operationalize this definition, a quadrat was counted as part of the home-range of a group if it was entered at least twice by one or more group members. This criterion of occupancy on at least two occasions was introduced to avoid including as "home-range" areas through which males, detached from their groups, occasionally moved; much of the area covered by males during the mating season was never used by the rest of the group. Since this study was primarily concerned with the habitual movements of whole groups, a definition of home-range that excluded this component was chosen.

Since an animal had only to be seen in a quadrat twice for all of it to be included in the home-range, the real area used by the group was probably overestimated. In both study areas the flat ground and homogeneity of vegetation provided few reference points and, during the first few months in each, only in the partly used peripheral quadrats was I able to map group movements on a scale finer than the 50 m grid. Later observations did suggest that there were small, unused lacunae within many quadrats that will have inflated estimates of total home-range size.

82

One cautionary note should be added concerning the comparison of home-range sizes of groups living in different habitats: it does not take into account differences in the use of vertical space. Although many studies, including this one, refer to *time spent* at different heights in the forest (Aldrich-Blake 1970; Chalmers 1968; Gartlan and Struhsaker 1972; Jolly 1966), only two dimensions are considered in estimates of home-range size. Yet Sussman (1974), for example, has shown important differences in the use made of the vertical component of a forest by *Lemur fulvus rufus* and *Lemur catta*. Two-dimensional comparisons alone of home-range size between species or between populations of a species occupying habitats with differing vertical components may thus be misleading.

b) *Size and differential use*

The study groups' home-ranges were between 6.75 and 8.50 ha (Table 4.7). This variation was not statistically significant, but in all cases ranges were more than twice the size of those reported by Jolly (1966) for *P. verreauxi* at Berenty.

TABLE 4.7

GROUP SIZE AND HOME-RANGE SIZE OF THE FOUR STUDY GROUPS, AND OF FIVE GROUPS AT BERENTY (DATA FOR BERENTY FROM JOLLY 1966)

LOCALITY	*Group*	*Group size*	*Home-Range size*
Northern	I	7	7.25 ha.
study area	II	4	8.50 ha.
Southern	III	6	6.75 ha.
study area	IV	4	7.00 ha.
Berenty	1		2.60 ha.
	2		2.60 ha.
	3		2.20 ha.
	4		2.20 ha.
	5		1.00 ha.

All four groups used some parts of their home-range much more heavily than others; all spent at least 60% of their time in only 10 quadrats and used a large part of their range rarely (Fig. 11). Kaufman (1962) used the term *core area* to describe " . . . a particular part of the home-range used more frequently and with greater regularity than other

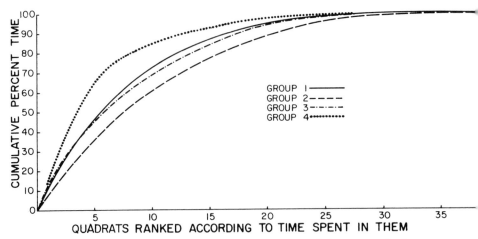

FIG. 11. Variation between each group's allocation of time across its home-range.

parts. . . . " The term was subsequently expanded to include the possibility of a home-range containing several core areas (Jay 1965; Mason
1968). In this study, *core area* was used to describe quadrats that, in
four or more months, were among those in which a group spent 75% of
its time when quadrats were ranked according to time spent in them.
Group I had six such quadrats, Group II seven, Group III ten, and
Group IV five (Fig. 12). There was a close correlation between quadrats
ranked according to this criterion and according to frequency of entry.

Core areas were not necessarily grouped together to form a composite
block, and although each group used these quadrats overall "more
frequently and with greater regularity than other parts," there was
considerable variation in time spent in them from month to month.
For example, Figure 13 shows that when Group II's five most-used
quadrats are considered month by month, there were some months
when one might not be occupied at all.

Although parts of each home-range could be differentiated according
to time spent in them, for Groups I, II, and III no differentiation within
these areas could be made according to activity. In Table 4.8, time
spent feeding/not-feeding by each group in core areas is expressed as
a percentage of total time spent feeding/not-feeding anywhere in the
home-range each month. It is apparent that for the three groups these

TABLE 4.8

MONTHLY ANALYSIS OF THE NATURE OF EACH GROUP'S ACTIVITIES IN ITS
AREAS OF INTENSIVE USE

		Time in foci of activity	
Group	Month	Feeding	Nonfeeding
I	July	53.3%	49.9%
	Aug.	19.7%	22.3%
	Oct.	38.3%	29.5%
	Nov.	42.0%	31.0%
	Dec.	53.2%	72.6%
	July	57.2%	56.0%
	Mean	43.9%	43.5%
II	July	38.4%	38.7%
	Aug.	48.0%	37.3%
	Oct.	33.8%	20.5%
	Nov.	38.2%	35.0%
	Dec.	36.1%	26.1%
	July	24.9%	33.2%
	Mean	36.5%	31.8%
III	Jan.	31.2%	31.1%
	Feb.	58.1%	75.8%
	Mar.	41.0%	37.3%
	Apr.	48.0%	53.3%
	May	29.5%	46.7%
	June	40.5%	79.8%
	Sept.	50.8%	57.3%
	Mean	42.7%	54.4%
IV	Jan.	39.4%	63.3%
	Feb.	43.3%	58.8%
	Mar.	61.2%	87.5%
	Apr.	54.0%	73.3%
	May	41.3%	69.3%
	June	49.8%	69.2%
	Mean	48.1%	70.2%

areas were not simply sleeping sites or areas of particularly intensive
feeding, but rather areas in which animals commonly both fed and
rested. Group IV spent more time in activities other than feeding in
their core areas (Mann-Whitney U Test, $U=1$, $p < .002$). The reason for
this is not known, but it is probably related to Group IV's attachment
to two clumps of *Alluaudia*, to one or other of which they almost
always resorted for the night. All groups had preferred sleeping trees

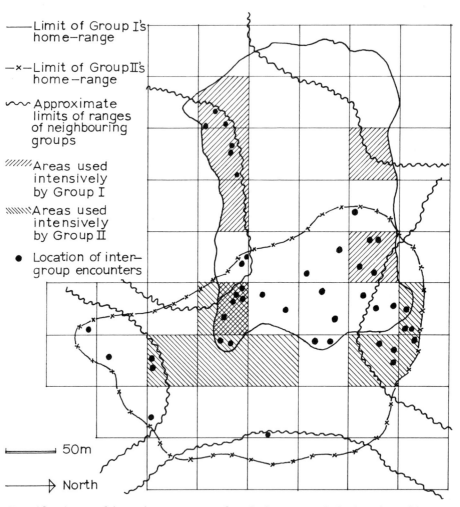

FIG. 12a. Areas of intensive use, areas of exclusive use, and the location of intergroup encounters within the home-range of the northern study groups.

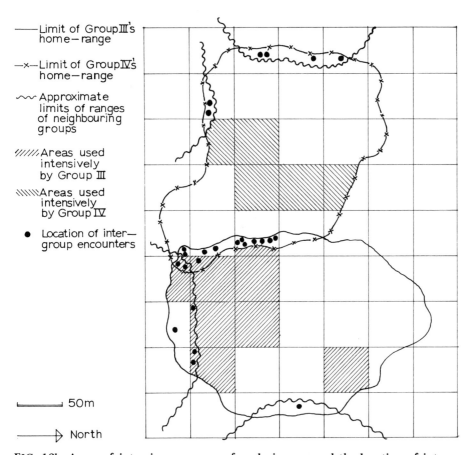

FIG. 12b. Areas of intensive use, areas of exclusive use, and the location of inter-
group encounters within the home-range of the southern study groups.

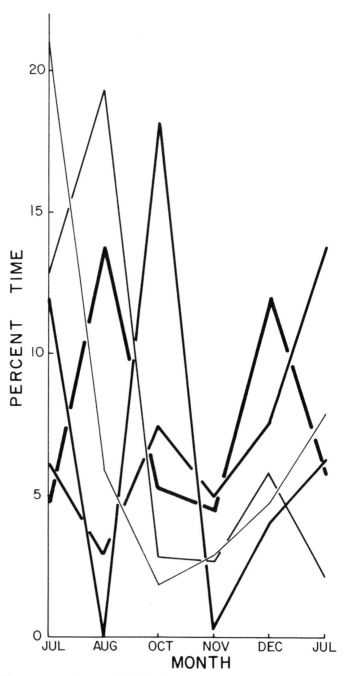

FIG. 13. Time spent each month in the five quadrats most used overall by Group II.

within some quadrats, but only Group IV returned to these trees almost every evening.

Seasonal changes in home-range utilization were marked. Considering data from the wet and dry season separately (Fig. 14), groups in both study areas spent more time in fewer quadrats in the dry season than in the wet. This tendency was more pronounced in the south than in the north.

In 1971 and 1974 Group III entered 18 different quadrats in the course of the June observation block. Of the 18 entered in June 1974, 14 were also entered in June 1971 and the remaining four quadrats were used by the group during other months. There was also a significant correlation in the way animals distributed their time between quadrats in the two months under consideration (Spearman Rank Correlation Coefficient, r_s -0.637, p.01). In both years, animals spent most of their time in a few parts of their range.

c) *Exclusivity of use*

The four study groups each had quadrats within their home-ranges where no other groups were seen, which were called *areas of exclusive use*. A quadrat was said to be an area of exlusive use if no other group was found in it more than once, and to be part of an overlap area if another group *was* found in it more than once. As with estimates of home-range size, this method probably resulted in some inaccuracies. Another bias may also be present in that I spent little time in quadrats used rarely by the study groups and thus was less likely to see other groups in these quadrats, some designated areas of exclusive use may have been areas in which I *did not see* other groups rather than areas into which other groups did not go.

In the north, Groups I and II had exclusive use of 46% and 43% of their total home-ranges. Intergroup interactions occurred throughout the extensive areas of overlap with neighboring groups, and there was no evidence that these encounters defined or defended the boundaries of a section of the home-range used exclusively by the resident group (Fig. 12). Exclusivity of use was not related to amount of use. Core areas were not necessarily areas of exclusive use and, conversely, little-used quadrats were not necessarily areas of overlap. Table 4.9 shows that many of each group's most- and least-used quadrats were also entered by other groups.

In the south, Groups III and IV had exclusive use of 87% and 91% of their total home-ranges (the observation period in 1974 was too short to permit a useful comparison). Intergroup encounters occurred along the periphery of this area of exclusive use and apparently served

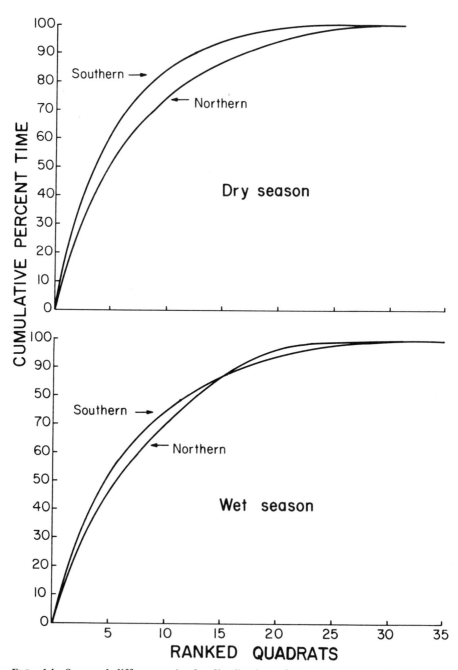

FIG. 14. Seasonal differences in the distribution of time between quadrats by the
groups in each study area.

TABLE 4.9

NUMBER OF QUADRATS USED BY OTHER GROUPS AMONG THE TEN LEAST- AND
THE TEN MOST-USED QUADRATS OF EACH STUDY GROUP

Group	*No. of quadrats in which overlap occurred among the 10 least-used quadrats in the home-range*	*No. of quadrats in which overlap occurred among the 10 most-used quadrats in the home-range*
I	7	8
II	10	10
III	2	3
IV	4	4

to define and/or defend its borders (see also chapter 7). In contrast
with the north, core areas almost all lay within the area of exclusive
use within each group's range.

2. Ranging Behavior

All four groups visited most parts of their home-range within ten
to twenty days. Characteristically, each group followed the same
general pattern and direction of movement for two or three days,
usually feeding in many of the same trees each day, and then suddenly
both the sleeping tree and the ranging pattern would change.

a) *Variation in daily distance moved*

The northern groups moved further each day than the southern
groups (Mann-Whitney U Test, U=34, p<.025). All four groups moved
further each day during the wet season than during the dry (Fig. 15).
(Mann-Whitney U Test, U=0, p<.001 — south; U=3, p<.008 — north.)
There was no significant difference in the distance moved by the
different age and sex classes within each group, nor between the groups
in each area. The data from each area were, therefore, combined.

In the north, the mean distance moved each day in the wet season
was 1,100 m; in the dry season it was 750 m. In the south, the mean
distance moved each day in the wet season was 1,000 m; in the dry
season it was 550 m.

Group III moved further, on average, each day in June 1974 than

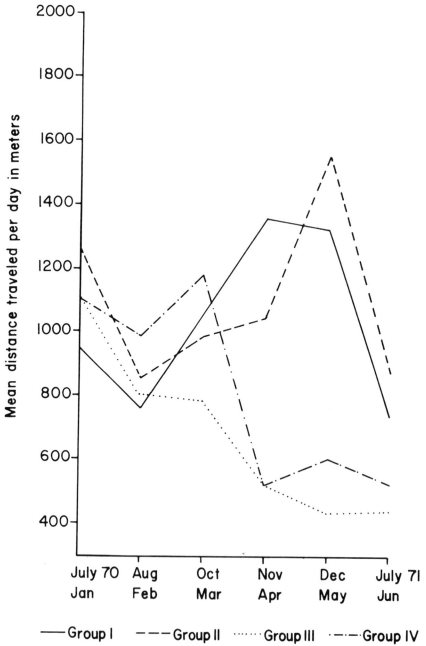

FIG. 15. Mean distance moved each day, each month, for each group.

in June 1971, the figures being 521 m and 451 m respectively (Mann-Whitney U Test, U=7, p<.04).

b) *Relationship between distance moved and area covered*

In all four groups, distance moved each day increased significantly during the wet season. At the same time, there was a slight increase in the number of different quadrats entered and a larger increase in the total number of quadrats entered (Table 4.10). This table also shows

TABLE 4.10

NUMBER OF QUADRATS ENTERED, NUMBER OF DIFFERENT QUAD-RATS ENTERED, AND THE RATIO BETWEEN THEM, FOR EACH GROUP FOR EACH MONTH

Group	Month	No. quadrats entered	No. different quadrats entered	Ratio
I	July	69	22	3.1
	Aug.	62	20	3.1
	Oct.	118	23	5.1
	Nov.	144	25	5.8
	Dec.	137	26	5.3
	July	69	17	4.0
II	July	90	24	3.7
	Aug.	79	21	3.8
	Oct.	122	23	5.3
	Nov.	121	26	4.6
	Dec.	167	27	6.2
	July	85	31	2.7
III	Jan.	119	27	4.4
	Feb.	88	23	3.8
	Mar.	95	19	4.0
	Apr.	69	20	3.4
	May	63	19	3.3
	June '71	55	18	3.0
	June '74	60	18	3.3
IV	Jan.	137	25	5.5
	Feb.	107	24	4.4
	Mar.	130	20	6.5
	Apr.	69	21	3.3
	May	56	18	3.1
	June	64	17	3.8

*Ratio = $\dfrac{\text{Total No. of Quadrats}}{\text{No. of different Quadrats}}$

the associated change in the ratio between these measures. The data suggest that, in ranging further in the wet season, groups tended to cover their home-range more rapidly in the wet season than in the dry, but that they did not expand it appreciably in the wet season. Despite differences in the mean daily distance traveled, the data on rate of quadrat use for June 1974 parallel closely those for June 1971.

It should be noted in association with these results that apparent home-range size increased with time spent watching each group faster when observations were initiated during the wet season than during the dry; increase in estimated home-range size was not simply a function of number of hours of observation, but of the season. This seasonal differential should be considered when comparisons of home-range size are made on the basis of short-term studies conducted in different seasons.

3. Feeding Behavior

An animal was recorded as feeding when it was ingesting or chewing bark, buds, leaves, flowers, or fruit. Animals detached food parts with their teeth, using their hands only to pull food-carrying branches to their mouths. They fed in many postures, enabling them to reach almost all parts of any tree. Although in the course of a day they used all substrate categories extensively except the ground, most feeding was done among smaller branches and twigs. Feeding, like other activities, occurred regularly at all levels in the forest; the two southern groups spent more time at lower levels than the northern groups, probably because of the lower stature of the southern forest (see 4(a) below and Richard, in press (a) and (b) for a fuller discussion of the physical parameters of feeding behavior).

a) *Time spent feeding*

In both study areas there were striking seasonal differences in the mean amount of time animals spent feeding each day (Fig. 16). Mean time spent feeding per day in the south during the wet season was 246 minutes (32.8% of total observation time), while in the dry season it was only 173 minutes (24% of total observation time). In the north, equivalent figures were 268 minutes (37%) and 245 minutes (30%). These results were calculated combining data from all age and sex classes (including gestating and lactating females), with the exception of data on the juvenile in Group III during the wet season. This animal was excluded from the analysis because he fed significantly longer each day in the wet season than did other animals in the two southern groups (Kruskal-Wallis, One-Way Analysis of Variance, $p < .01$). Combining data

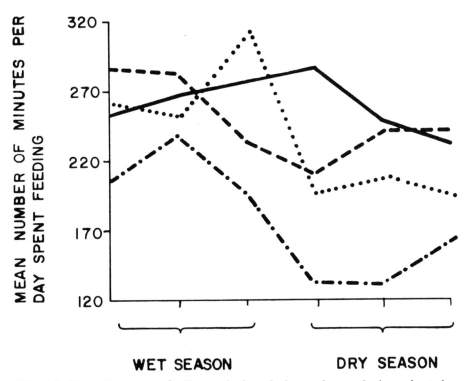

FIG. 16. Mean time spent feeding each day, during each month, in each study group.

for the two groups in the south, the amount of time spent feeding each day in the wet season was significantly greater than in the dry season (Mann-Whitney U Test, p<.004). In the north, the difference was not significant, although there was a similar trend. Comparing results from the north and south for each season, there was no significant difference during the wet season, but in the dry season the southern groups fed significantly less than those in the north (Mann-Whitney U Test, p<.001).

Just as animals in Group III moved further each day in June 1974 than in June 1971, so they fed longer; they spent a mean of 29.8% (214.5 minutes) of total time each day feeding, as compared with 27% (198 minutes) in 1971 (Mann-Whitney U Test, U=4, p <.05).

In addition to seasonal variation in time spent feeding, there appeared to be considerable variation in the rate at which food was consumed within a given feeding bout; dead wood, for example, had to be "chiseled out" laboriously, whereas leaves could be steadily gathered in, ingested, and chewed. At some point during most bouts, all animals could be found in a single tree feeding almost without interruption for periods of up to two hours. For the rest of the bout, animals would disperse and move rapidly from tree to tree, feeding slowly in each. These differences were not quantified, but underline the inaccuracy of equating time spent feeding with quantity of food eaten.

b) *Species composition of diet*

  i) Regional variation

The species composition of the diet of the groups in each study area was almost totally different. Species that each group spent most time eating throughout the study are listed in Table 4.11. The regional variation was largely, but not entirely, due to differences in the composition of the two forests. Four tree species were found in both forests. *Baudouinia fluggeiformis* (found in 92% of vegetation samples in the north and 23% in the south) was an important dietary component in the north but untouched in the south. *Cedrelopsis grevei* (17% in the north, and 77% in the south) was eaten commonly in the south but never in the north. *Commiphora pervilleana* (66% in the north, and 77% in the south) was eaten commonly in the north, but rarely in the south. Finally, *Rothmannia decaryi* (79% in the north and 3% in the south) was eaten in quantity in the south, and rarely in the north; further, in the south animals ate only the green ripe fruit of this species, whereas in the north animals ate only its large white flowers.

TABLE 4.11

FOOD SPECIES EATEN BY EACH GROUP, RANKED ACCORDING TO TIME SPENT
FEEDING ON EACH (SPECIES EATEN FOR LESS THAN 1% OF TOTAL TIME SPENT
FEEDING NOT INCLUDED)

Group I		Rank	Group II	
Food species	Time spent*		Food species	Time spent*
Drypetes sp. No. 18	12.4	1	Drypetes sp. No. 18	11.7
Cedrelopsis sp. No. 471	5.4	2	Cedrelopsis sp. No. 471	8.9
Liana No. 215	5.3	3	Liana No. 13	4.5
Dead Wood	5.3	4	Commiphora pervilleana	4.4
Capurodendron microlobum	5.1	5	Bathiorhamnus louveli	3.9
Rheedia arenicola Jerm & Perr	4.5	6	Dead wood	3.9
Commiphora pervilleana	4.2	7	Rhopalocarpus similis	3.7
Liana No. 312	4.0	8	Liana No. 215	3.6
Liana No. 38	3.7	9	Liana No. 38	3.3
Protorhus deflexa	3.4	10	Boscia sp. No. 302	3.1
Liana No. 36	3.3	11	Protorhus deflexa	2.4
Mundulea sp. No. 64	2.9	12	Capurodendron microlobum	2.3
Liana sp. No. 13	2.2	13	Mammea sp. No. 5	2.2
Polyalthia sp. No. 116	2.1	14	Liana sp. No. 36	2.2
Cedrelopsis sp. No. 123	1.5	15	Rheedia arenicola Jerm & Perr	2.1

*Results are expressed as a percentage of total time spent feeding.

TABLE 4.11: Cont.

Group I		Rank	Group II	
Food species	Time spent*		Food species	Time spent*
Baudouinia fluggeiformis	1.5	16	Machersonia gracilis	1.9
Erythroxylon sp. No. 514	1.5	17	Baudouinia fluggeiformis	1.8
Sp. No. 433	1.5	18	Boscia sp. No. 301	1.6
Boscia sp. No. 301	1.5	19	Mimusops sp. No. 320	1.6
Sp. No. 211	1.4	20	Liana sp. No. 296	1.6
Machersonia gracilis	1.1	21	Holmskioldia microcalyx	1.3
Holmskioldia microcalyx	1.1	22	Sp. No. 634	1.2
Liana No. 452	1.1	23	Grewia sp. No. 121	1.2
Mammea sp. No. 5	1.1	24	Malleastrum sp. No. 240	1.0
Boscia sp. No. 302	1.1	25	— —	
Boscia sp. No. 603	1.0	26	— —	
Liana sp. No. 042	12.3	3	Mimosa sp. No. 033	11.2
Grewia sp. No. 089	7.7	4	Grewia sp. No. 059	9.9
Terminalia sp. No. 09	5.0	5	Hazunta modesta	4.8
Diospyros humbertii	4.1	6	Terminalia sp. No. 09	4.3
Grewia sp. No. 059	5.2	7	Grewia sp. No. 089	2.6

*Results are expressed as a percentage of total time spent feeding.

TABLE 4.11: Cont.

| Group I | | Rank | Group II | |
Food species	Time spent*		Food species	Time spent*
Euphorbia plagiantha	2.3	8	Liana sp. No. 053	2.5
Liana sp. No. 054	2.1	9	Diospyros humbertii	1.9
Liana sp. No. 053	1.9	10	Rothmannia decaryi	1.6
Hazunta modesta	1.8	11	Commiphora sp. No. 092	1.6
Liana sp. No. 056	1.7	12	Grewia sp. No. 0126	1.5
Sp. No. 0113	1.4	13	Liana sp. No. 0125	1.5
Liana sp. No. 058	1.3	14	Entada abyssinicus	1.4
Cedrelopsis grevei	1.3	15	Liana sp. No. 054	1.4
Operculicarya decaryi	1.2	16	Cedrelopsis grevei	1.0
Liana sp. No. 0146	1.1	17	Albizia sp. No. 034	1.0
— —		18	Commiphora sp. No. 076	1.0

Group III			Group IV	
Terminalia sp. No. 048	21.5	1	Terminalia sp. No. 048	21.0
Mimosa sp. No. 033	16.8	2	Liana sp. No. 042	15.4

*Results are expressed as a percentage of total time spent feeding.

ii) Local variation

Some variation was seen between the diets of the two neighboring study groups in each area. In the north Group I ate one species, *Malleastrum* sp., on which Group II was never seen to feed, and Group II ate one,

Erythroxylon sp., never eaten by Group I. In the south all food species eaten by one group were also eaten by the other.

In both areas, eight out of the twelve species eaten most commonly by each group were the same, and there was a close correlation between the way the two groups in each area allotted feeding time to these species (Spearman Rank Correlation Coefficient $p < .01$ — north and south).

There was more local variation in the south than in the north. Of the food species eaten for more than 1% each of total time spent feeding by Group III, 41% were eaten for less than 1% of total time spent feeding by Group IV, and of the food species eaten for more than 1% each of total time spent feeding by Group IV, 33% were eaten for less than 1% each of total time spent feeding by Group III. Equivalent values for Groups I and II were 23% and 33%.

iii) Seasonal variation

The composition of each group's diet changed almost completely between seasons. Only five food species out of the observed total of 77 were eaten by the southern study groups for more than 1% of total time spent feeding in both dry and wet seasons. The northern groups ate 9 out of 99. Figure 17 shows the percentage of total feeding time per month spent by Groups I and III feeding on the two food species that were most commonly eaten by each group in the course of the study, and demonstrates two distinct patterns of utilization. In both groups, one species — *Drypetes* sp. in the north and *Mimosa* sp. in the south — was an important dietary item throughout the study; in contrast, the second species — *Cedrelopsis* sp. in the north and *Terminalia* sp. in the south — constituted the bulk of the group's diet during a short period, apart from which it was of little or no importance.

iv) Long-term variation

The 1974 follow-up study indicated considerable long-term stability in food preferences. Table 4.12 lists food species eaten for more than 1% of total feeding time by Group III in June of each year, and shows some striking similarities; the most eaten species in both years, for example, belonged to the genus *Mimosa* and was fed on for 20.1% of total feeding time in 1971 and 20% in 1974. In fact there was a close overall correlation between the way in which animals distributed their feeding time among species in the two years (Spearman Rank Correlation Coefficient, $r_s = 0.93647$, $p < .01$).

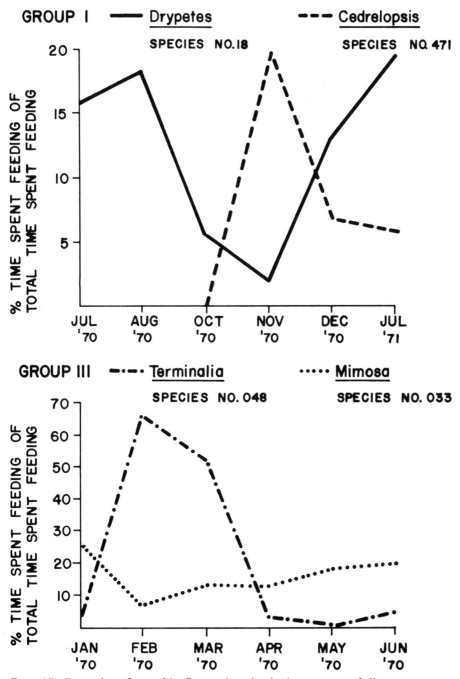

FIG. 17. Examples of monthly fluctuations in the importance of dietary components.

TABLE 4.12

FOOD SPECIES EATEN BY GROUP III FOR MORE THAN 1% OF TOTAL TIME SPENT
FEEDING IN JUNE 1971 AND JUNE 1974

Family	Species	Vernacular name	1971 % feeding time	Rank	1974 % feeding time	Rank
Mimosaceae	Mimosa sp. 033	kirava	20.1	1	20	1
Tiliaceae	Grewia sp. 089	hazovonge	16.4	2	–	–
Combretaceae	Terminalia sp. 09	taliforokoka	13.8	3	13	3
Convolvulaceae	Ipomaea sp. 0146	baha	5.0	4	5	7
Convolvulaceae	Terminalia sp. 048	fatra	4.5	5	3	9
– – –	Sp. No. 0113	– – –	4.2	6.5	3	9
Papilionaceae	Rhynchosia Sp. No. 042	tsarafangitsa	4.2	6.5	–	–
Tiliaceae	Grewia sp. 059	tabarika	4.0	8.5	–	–
Papilionaceae	Clitoria heterophylla Lam.	kelelo	4.0	8.5	3	9
Anacardiaceae	Operculicarya decaryi H. Perr	zaby	3.8	10	2	11
Ebenaceae	Diospyros humbertiana H. Perr	maintefo	2.5	11	5	6
Rubiaceae	Paederia sp. 058	korokatsa	2.4	12	–	–
– – –	Other	– – –	2.3	13.5	2	13
Euphorbiaceae	Euphorbia plagiantha	fihanga	2.3	13.5	9	4
Meliaceae	Cedrelopsis grevei H. Bail	katrafay	2.1	15	18	2

TABLE 4.12: Cont.

Family	Species	Vernacular name	1971 % feeding time	Rank	1974 % feeding time	Rank
— — —	Liana sp. 053	tamboro	2.0	16	—	—
Loganiaceae	*Strychnos* sp. 012	dagoa	1.6	17	—	—
— — —	Liana sp. 055	vaheranga	—	—	7	5
Asdepiadaceae	*Marsdenia brevisquamma* Jerm. et Perr	saryfitovy	—	—	2	13
— — —	*Poivrea* sp. No. 028	— — —	—	—	2	13
Burseraceae	*Commiphora* sp. 076	darotandroka	—	—	1	15.5
Cucurbitaceae	*Corollocarpus grevei* keraudren	lelantrandraky	—	—	1	15.5

c) *Species diversity*

i) Regional variation

The northern groups' diet contained more species than that of the southern groups. In six months of observation, Group I was seen feeding on 85 different food species (including 18 liana species), and Group II on 98 (including 18 liana species). In the south, Group III was seen to eat 77 different food species (including 14 lianas) and Group IV 65 (including 13 lianas). These represented 34.5% (Group I), 41% (Group II), 64% (Group III) and 52.5% (Group IV) of all tree species identified in each forest.

In both areas many food species were eaten rarely. Figure 18 shows that all four groups spent at least 60% of total feeding time eating only 12 food species; it also indicates that this tendency to feed primarily on a few species was more pronounced in the south than in the north (Kolmogorov-Smirnov Two-sample Test, N=12, K_d=29.1 p<.01).

ii) Seasonal variation

The number of food species eaten by each group declined in the wet season in both study areas. When only the number of food species on

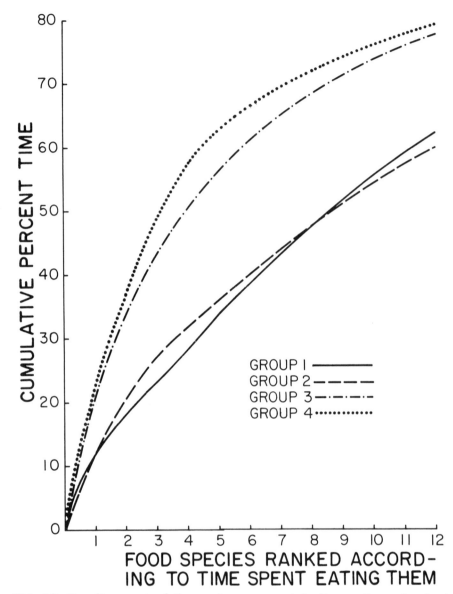

FIG. 18. Overall amount of time each group spent feeding on the twelve food species eaten by each group more commonly than any other food species.

which each group spent more than 1% of total feeding time are compared (Table 4.13), the difference between seasons is statistically signif-

TABLE 4.13

NUMBER OF FOOD SPECIES ON WHICH EACH GROUP SPENT MORE THAN 1% OF ITS TOTAL TIME SPENT FEEDING, IN EACH SEASON

Month	*No. of species*		*Month*	*No. of species*	
	Group I	*Group II*		*Group III*	*Group IV*
July & August	18	22	May & June	18	19
Nov. & Dec.	13	17	Jan. & Feb.	8	6

icant for all groups (Mann-Whitney U Test, p<.01). (When all food species were included in the analysis, this trend was masked in the north by the long tail of rarely eaten food species. This is not, however, to dismiss the possible importance of such foods.)

iii) Long-term variation

The number of species eaten by Group III was almost the same in June 1974 as in June 1971; the figures were 25 and 29 respectively. Further, over 1% of total feeding time was spent eating 16 species in 1974 and 17 species in 1971.

d) *Food parts eaten*

i) Regional variation

Figure 19 shows that there were marked differences in the overall quantity of various food parts eaten by the groups in each study area. Animals ate more fruit and mature leaves in the south than in the north, where immature leaves constituted an important dietary item.

Comparing data from the two study areas by season, further distinctions emerge. In both seasons the southern groups ate more mature leaves and fewer immature leaves than the northern groups (Mann-Whitney U Test, p<.01), but only in the wet season did they eat more fruit than in the north (Mann-Whitney U Test, p<.008). In neither season was there a regional difference in the time spent eating flowers.

ii) Seasonal variation

The nature of seasonal changes in food parts eaten was similar in the two groups in each study area and, in both forests, reflected closely changes in the availability of food parts. Figure 20 shows the percentage of total feeding time that the northern groups spent eating different

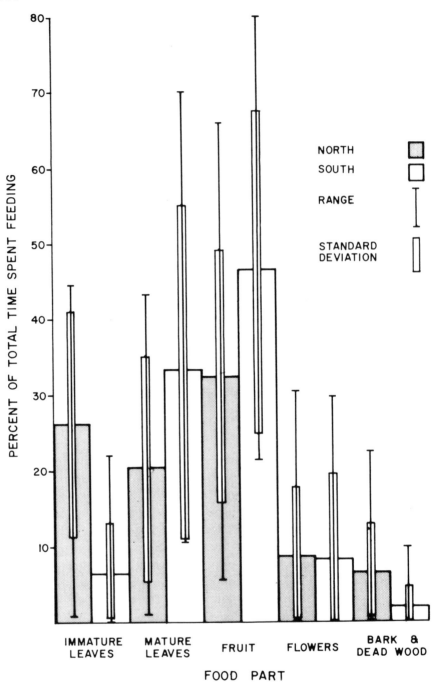

FIG. 19. Amount of time spent eating different food parts, irrespective of season, by the groups in each study area.

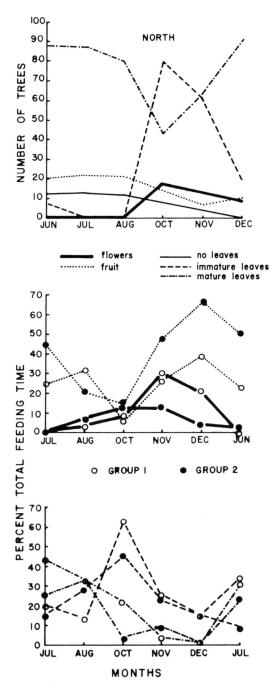

FIG. 20. Percentage of total feeding time each month that the northern groups spent eating different food parts, and fluctuations in the abundance of those food parts.

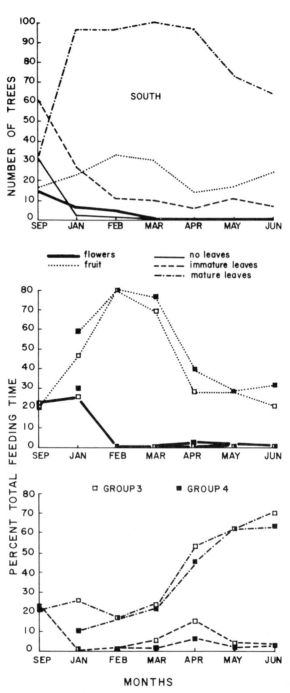

FIG. 21. Percentage of total feeding time each month that the southern groups spent eating different food parts, and fluctuations in the abundance of those food parts.

food parts each month together with the information from Figure 10 concerning fluctuations in the abundance of these parts. Figure 21 shows similar data for the southern groups.

In the north, a general peak in immature leaf consumption occurred at the same time as a decrease in mature leaf consumption at the beginning of the wet season. Further into the wet season, consumption of immature leaves declined too, and there was an increase in the consumption of flowers in November, followed by a large increase in the time spent feeding on fruit in December. In the dry season, there was a general decrease in the time spent feeding on fruit and flowers, and a general increase in time spent feeding on mature leaves. Time spent feeding on immature leaves decreased generally in the dry season, although the figures for Group I in July 1971 did not conform to this pattern. A large proportion of the time this group spent feeding on immature leaves was spent in a single tree of *Capurodendron microlobum,* which was covered with leaf buds at that time. Phenological data indicate that at that time budding occurred uniquely in this one tree.

In the south flowers were an important dietary item at the onset of the wet season in January but later, as in the north, fruit constituted an overwhelming proportion of the animals' diet. At the beginning of the dry season, as in the north there was a general increase in the amount of time spent feeding on mature leaves, and this trend continued through the dry season until the end of the study in June. Flowers again became an important dietary component toward the end of the dry season in Group III (no observations were made on Group IV). These flowers came uniquely from *Alluaudia ascendens* and *A. procera.* Both these species were abundant in the forest, and both flower annually in September. September was also the only month in which immature leaves constituted an appreciable proportion of the diet. This peak came shortly after light, but unexpected and atypical rainfall at the beginning of September 1970, which was sufficient to stimulate leaf production by many species; it was this sudden availability of immature leaves that was reflected in Group III's diet at that time.

iii) Long-term variation

The percentage of total feeding time spent eating fruit, flowers, immature leaves, mature leaves, and bark by Group III in June 1971 and June 1974 is shown in Table 4.14. In 1971 fruit of a *Grewia* sp. constituted 21.8% of the animals' diet. In 1974 they ate almost no fruit and instead spent significant amounts of time feeding on the bark of *Cedrelopsis grevei*; the percentage of mature leaves in their diet was also

TABLE 4.14

PERCENTAGE OF TOTAL FEEDING TIME SPENT EATING DIFFERENT FOOD PARTS
BY GROUP III IN JUNE 1971 AND JUNE 1974

	Fruit	Flowers	Immature Leaves	Mature Leaves	Bark	Other (unidentified)
June '71	21.8	0.6	3.8	70.8	0.8	2.6
June '74	0.1	—	3.5	85.9	8.5	2.0

Total minutes spent feeding: 1971 — 1235 mins.
1974 — 1284 mins.

higher. In both years, flowers and immature leaves formed a negligible portion of their diet at that time, and in both, mature leaves were easily the single most important item.

iv) Bark and dead wood as dietary components

Groups fed extensively on bark and/or dead wood at certain times of year (Fig. 22). Dead wood was never eaten in the south, but bark formed an important dietary item during the dry season. In September 1970 Group III spent 15.5% of total feeding time eating the bark and cambium of *Operculicarya decaryi*. This wood contains 81% water by weight. Animals gouged out the thin-barked, soft, moist wood with their "toothcombs," leaving scars on the trunk up to 1 cm. deep and 4 cm. long. In June 1971 animals again began to eat this wood. In June 1974 Group III spent prolonged periods eating the bark of *C. grevei*. The water content of this wood was not estimated but was evidently no higher than in many other species. It is interesting to note that this bark is extensively used for medicinal purposes by the Malagasy.

In the north, both groups ate bark to the exclusion of dead wood in the dry season, and dead wood almost to the exclusion of bark in the wet season. The bark eaten in the dry season came mainly from thin branches of *Commiphora pervilleana;* the bark alone was stipped off, using first the "tooth-comb" to prise up the bark and then the premolars to tear it off. The underlying wood was not gouged out as in the south and the bark appeared to contain little water.

In the wet season, dead wood was a major item in the diet of both Groups I and II. Each group would daily cluster round a dead tree trunk and tear off splinters of wood with their "tooth-combs" and premolars. Only two hunks of dead wood were used by each group in this way, although other trunks were present in the home-ranges of both that did not differ noticeably from those used as food sources; the

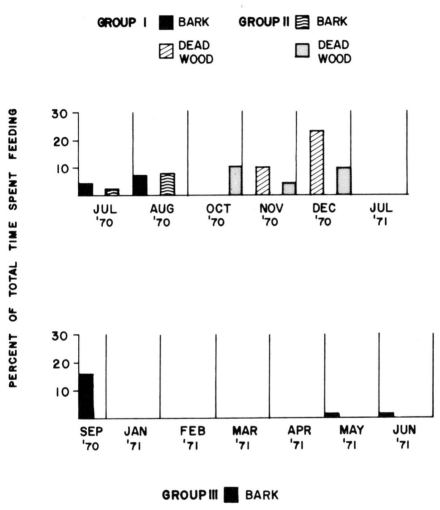

FIG. 22. Percentage of total feeding time each month that each group spent eating bark and dead wood.

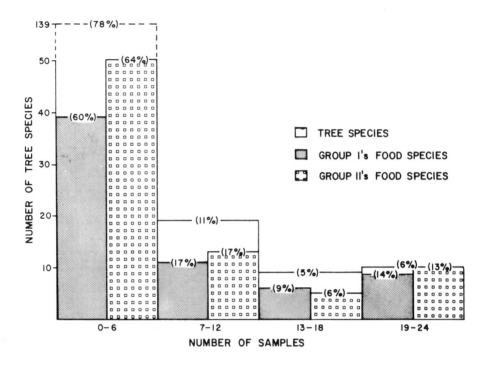

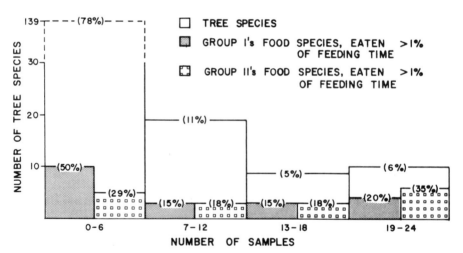

FIG. 23. Distribution of food species in the north between tree species-frequency categories.

latter were very dry, very hard, and very dense. Close examination of the wood did not reveal any signs of bore holes, so it is unlikely that animals were gnawing through the wood to reach insect eggs or larvae.

e) Diet and the abundance of food

Relative to the differing abundance of tree species in each forest, groups used as food a high number of species that occurred frequently in the forest, instead of seeking out more rarely occuring species. In Figure 23, tree species identified in the vegetation analysis in the north are grouped according to the frequency with which they were present in samples. Four categories are shown: rare species (present in 1-6 samples, or identified only because animals ate them), moderately rare species (present in 7-12 samples), moderately abundant species (present in 13-18 samples), and abundant species (present in 19-24 samples). Lianas are excluded from the analysis because no estimate was made of the abundance of individual species. In Figure 23(a), the total number of food species eaten by the northern groups in each of these categories is shown. In Figure 23(b), only preferred food species — that is, foods eaten for more than 1% of total feeding time — are included. While rare species formed a major component (60% for Group I and 64% for Group II) of the total number of species in the animals' diet, over half the total number of rare species present in the forest were untouched by either group. In contrast, although abundant species constituted a small proportion (14% for Group I and 13% for Group II) of the total species number in the animals' diet, almost all abundant species present in the forest were eaten. When only preferred foods are considered, the number of abundant species is higher (20% and 30%). In other words, these few abundant species were almost all eaten for extended periods.

Figure 24 shows comparable results for the southern study area. As in the north, animals fed on almost all of the few species that were abundant in the forest, although these constituted a small proportion (7% for Group III and 6% for Group IV) of the total number of species in their diet; over half the species eaten by each group were rare. In contrast to the north, animals ate a much higher proportion of the rare species present in the forest, but it should again be noted that the diversity of the southern forest was lower than that of the north. When only preferred foods are considered, an unexplained difference between Groups III and IV emerges. The distribution of Group III's foods between categories is similar to that of the northern groups; almost all the abundant species in the forest were included among their preferred foods and indeed constituted close to one-third of them. In contrast,

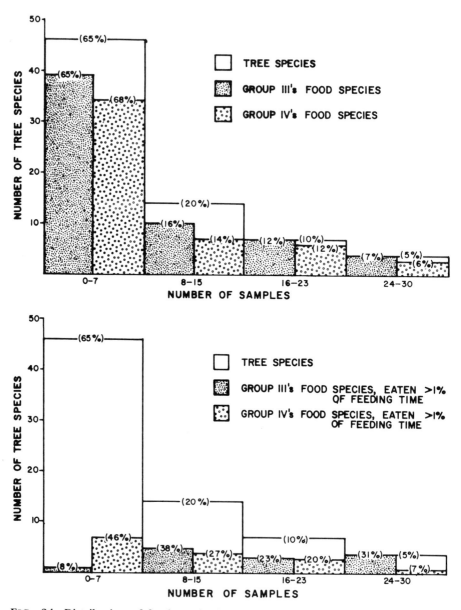

FIG. 24. Distribution of food species in the south between tree species-frequency categories.

only one of Group IV's preferred foods was abundant in the forest, and about half of them were rare.

In summary, three of the four groups ate each of almost all the most abundant tree species in the forest for significant amounts of time. Their diet was supplemented by a wide variety of rarer species, only some of which were eaten commonly. In the north, many rare species were untouched by either group, whereas in the south, where overall species diversity was lower, animals fed on most of these rare species too. In Group IV, animals ate many rare but few abundant species.

4. Positional Behavior

a) Posture, height above ground, and substrate use
The title *Vertical Clinging and Leaping* (Napier and Walker 1967) aptly describes *P. verreauxi's* saltatory movement between vertical trunks when moving through the forest from one food source to another or when fleeing (Plate X); but animals could also adopt a wide range

PLATE X. **Adult female leaps from the trunk to which she has been clinging**

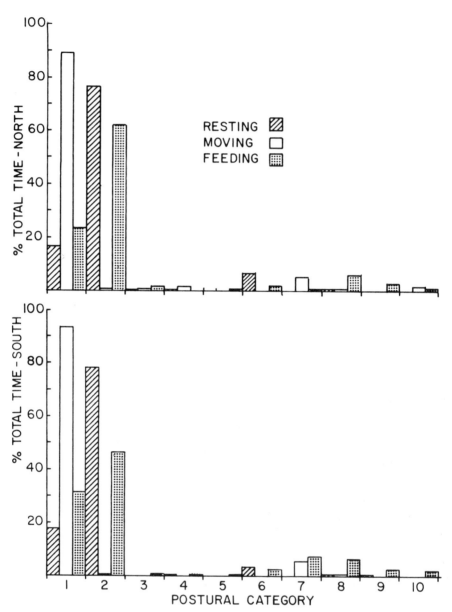

FIG. 25. Time spent in different postures while resting, feeding, and moving by the study groups in each area.

of postures. Figure 25 shows how much time the four groups spent in different postures while resting, feeding, and moving (see chapter 2 for description of postural categories). There was a striking similarity in the results for each study area. When moving, animals in both areas remained almost exclusively upright; they generally sat to rest. In both study areas, over 75% of feeding was done sitting or clinging vertically, but it was also in this activity that the greatest diversity of posture was shown; through adopting a variety of postures, animals could reach food located on the periphery of trees.

Eight height categories were established in each forest: ground, off-ground-1 1/2 m, 1 1/2m-3 m, 3 m-5 m, 5 m-7 m, 7 m-10 m, 10 m-13 m, and over 13 m. Significant differences occurred in the distribution of time between these categories by the animals, according to activity, season, and study area (Figs. 26 and 27). In the north, there was little difference within each season between the time spent at different heights while feeding and while engaged in activities other than feeding (Spearman Rank Correlation Coefficient, $r_s=.929$, $p<.01$, $r_s=.992$, $p<.01$). However, there was a marked difference in feeding heights between the two seasons: during the dry season, animals spent more time feeding high in the forest, often on the buds and leaves of emergent trees. During the wet season, they tended to concentrate their feeding activity in the dense and more or less continuous vegetation between 3 m and 10 m (Fig. 26).

In the south there was little difference during the wet season between the time spent at different heights while feeding and while engaged in activities other than feeding (Spearman Rank Correlation Coefficient, $r_s=.929$, $p<.01$). In the dry season, animals spent significant periods high in the forest engaged in activities other than feeding; this reflects the increase in the amount of time spent sunning in the tree tops early in the morning during this season. Unlike the north, feeding heights were similar in the wet and dry seasons (Spearman Rank Correlation Coefficient, $r_s=.97$, $p<.01$). Time spent at different heights in activities other than feeding did differ between seasons (Fig. 27), probably as a result of the increased time spent sunning in the tree tops by animals during the cooler dry season.

Although animals in both study areas used all the height categories available to them to some extent in the course of observations, there was no significant correlation between the two study areas in the time spent in each by animals, whether feeding or in activities other than feeding. This difference was probably associated with differences in the height and structure of the two forests.

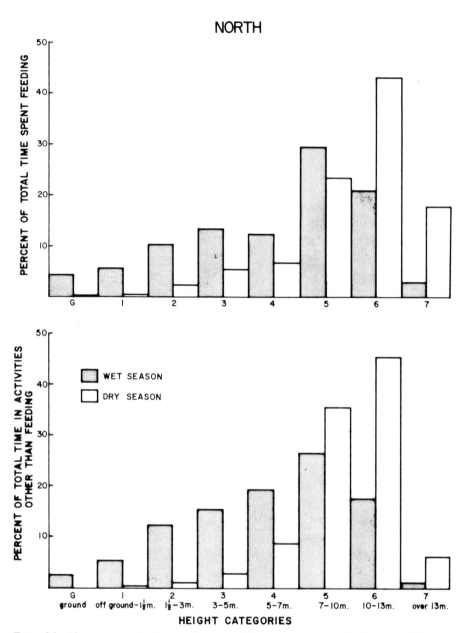

FIG. 26. Time spent feeding and in activities other than feeding at different
heights in the forest, in each season, by the northern groups.

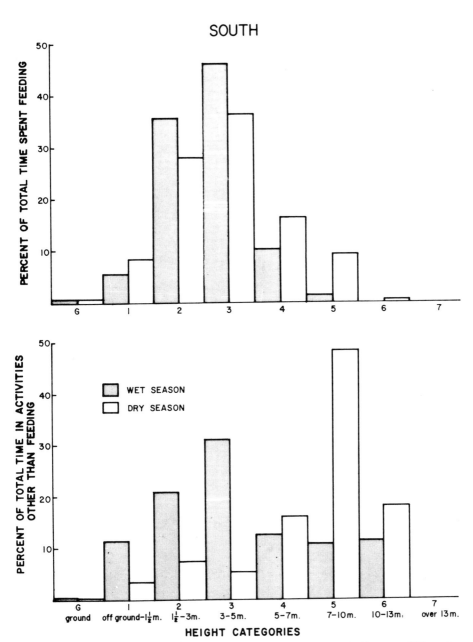

FIG. 27. Time spent feeding and in activities other than feeding at different heights in the forest, in each season, by the southern groups.

Five substrate categories were defined for this study: Ground, Main Vertical, Main Horizontal, Smaller Branches (less than 6 cm in diameter) and Twigs (less than 1 1/2 cm in diameter). Although animals in both areas spent little time on the ground, all other substrate categories were used extensively (Fig. 28). Indeed, the similarity of the results in the

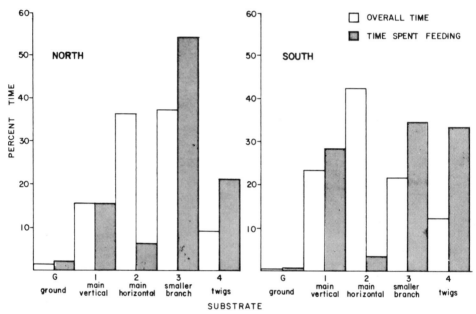

FIG. 28. **Time spent on different substrates by the northern and southern groups overall and when feeding.**

two study areas contrasts strikingly with the considerable differences apparent between the physical structure of the two forests (2B.2). Further, when distribution of time between substrate types during feeding alone is considered, the same trends can be seen in the data from both areas (Fig. 28); animals tended to feed from smaller branches and twigs, rather than from main trunks and branches.

b) Vertical clinging and leaping as an adaptation

Much has been written concerning the adaptiveness of *P. verreauxi's* locomotor pattern. In 1962 (b), Petter commented that a "... leaping type of locomotion is well fitted to predator avoidance...," but that vertical clingers and leapers in general are confined to vertical branches and trunks and at a disadvantage on small or horizontal branches; he saw this as the basis for the ecological separation of Indriids from sympatric Lemurids. Jolly (1966) did not totally agree with this: ". . . *Propithecus* can and does exploit whatever branch type is available, although it frequently moves and rests on vertical branches." On the subject of time spent at different heights above the ground she endorsed Petter's (1962b) impression: ". . . *Propithecus* and *Lemur* keep to much the same heights when in the trees, although *L. catta* descends more frequently to the ground."

Napier and Walker's (1967) category *Vertical Clinging and Leaping* was coined to describe a locomotor pattern in which ". . . the body is held vertically at rest and pressed to the trunk or main branch of a tree; movement from place to place is effected by a leap or jump from one vertical support to another" (Napier and Napier 1967). A number of prosimians, both extant and extinct, were allocated to this category on morphological grounds. In discussing the adaptiveness of this type of locomotion, Napier and Walker (1967) proposed that it evolved as a predator-avoidance mechanism; since there are no large arboreal predators in Madagascar today, they added, the behavior must be a "leftover" from late Eocene times. They also postulated that the behavior permits ". . . the use of certain restricted habitats where only vertical supports are found, such as the *Alluaudia* scrub of S.W. Madagascar. . .," but that ". . . the limitations imposed by the vertical clinging and leaping habit upon feeding behavior are also quite marked; the large forms, especially, are at a disadvantage when feeding in a small branch milieu."

Cartmill (1972) dismisses this view of vertical clinging and leaping as a kind of locomotor anachronism in Madagascar. He notes *P. verreauxi*'s alarm calls, suggesting these indicate continued predator pressure, and he quotes Jolly's (1966) comments on the locomotor versatility of the animals she studied at Berenty. His conclusion is that the leaping form of locomotion shown by *Propithecus* may be both a predator-avoidance mechanism and an adaptation for feeding on leaves in higher levels of the forest. He quotes in support of this Napier's (1963) statement that canopy-feeding primates ". . . must, in moving through the forest, either descend to lower levels where the canopy is continuous or pass from crown to crown in the discontinuous stratum by leaping or brachiating."

Data from this study endorse Jolly's findings and support Cartmill's (1972) conclusions; alarm calls given against raptors were found to be accompanied by a vertical-substrate seeking locomotor response, and there was no evidence that locomotor limitations prevented *P. verreauxi* from reaching all parts of the tree accessible to sympatric quadrupeds

PLATE XI. Adult female with infant makes her way up an *Alluaudia ascendens* branch

such as *L. fulvus*. Furthermore, results from both this and Jolly's study support Cartmill's hypothesis that vertical clinging and leaping facilitates the exploitation of food sources in the emergent layer of the forest; although the southern forest did not have an "emergent layer," *P. verreauxi* did exploit seasonally available flowers at the tips of tall, vertical *Aluaudia* branches (Plate XI). In the north, animals spent significant amounts of time, particularly in the dry season, feeding in emergent trees. Descriptive notes made in 1974 at Antserananomby where *P. verreauxi* is sympatric with *L. catta* and *L. fulvus* strongly suggested that *P. verreauxi* alone of the three species fed for appreciable periods in emergent trees in the forest; *L. fulvus* lives almost exclusively in the continuous canopy, and *L. catta*'s vertical range includes all levels from the ground to the continuous canopy, but rarely emergents (Sussman 1974). Jolly's (1966) data support this conclusion. It is of interest that a pattern of high forest exploitation also appears to be present in the other large Indriid, *Indri indri* (Pollock pers. comm.)

5. Resting

An inactive animal was said to be resting. This included both sunning behavior and true resting, because no criteria were found to discriminate consistently between them. Distinctions were possible only at a general descriptive level.

During "true" rest periods animals, singly or in groups of two or more, sat or sprawled, usually in the shade. On hotter days in both study areas, animals often moved down to a height of 2-3 m or even to the ground to rest. On cooler days groups preferred wide, shaded, horizontal branches or, in the south, the fork of an *Alluaudia* tree. As a result of this choice of substrate, animals were usually at least 4 m from the ground. Self- and allo-grooming and, in the wet season, play bouts took place during these rest periods.

Sunning behavior (Jolly 1966) occurred after sunrise in the dry season in both study areas before any major activity took place, and sometimes again before the animals settled for the night. On most mornings in the dry season, animals moved on to the first main branches to be struck by the sun and stayed there for about one hour. Posture varied during these sunning bouts from the habitual sitting position, with any part of the animal turned to the sun, to a position in which the animal sat with arms and legs splayed, its head on one side, and its black-skinned ventral surface, only sparsely covered with fur, facing the sun (Plate XII). Such behavior was less frequent in the wet season. Similar

PLATE XII. Male sunning in the early morning

accounts of sunning have been given for *Lemur catta* and *Lemur fulvus rufus* (Jolly 1966; Sussman pers. com.).

6. Daily Patterns of Activity

Figures 29 and 30 show the mean number of minute records per hour on which subjects were recorded feeding and resting, respectively, throughout the day; the range of variation is also shown. The daily data for each area were grouped into wet and dry season means, which are plotted separately. Almost all movement occurred in association with feeding bouts or travel to a preferred sleeping tree at the end of the afternoon feeding bout, so these data were not included.

a) Feeding

In both areas, the pattern of feeding in each season and the nature of the change between seasons were clear and similar; during the wet season, feeding began early, reaching a peak between 0700 h and 0900 h.

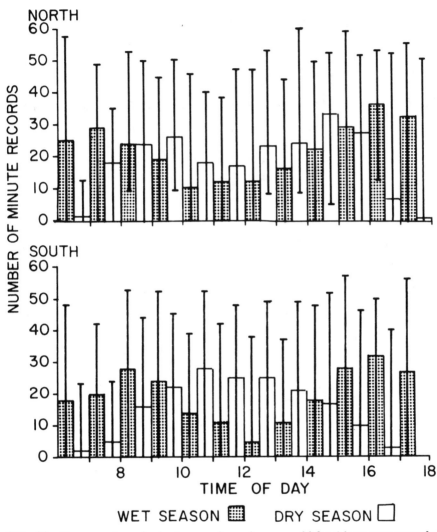

FIG. 29. Mean number of minute records per hour on which subjects were recorded feeding throughout the day, in each season and study area.

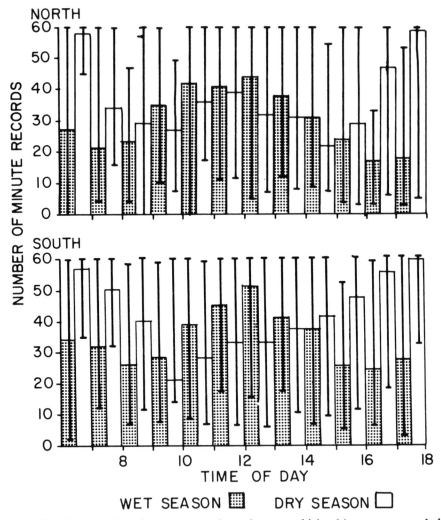

FIG. 30. Mean number of minute records per hour on which subjects were recorded resting throughout the day, showing seasonal changes.

This was followed by a gradual decrease until midday, by which time very little feeding was taking place. Between 1300 h and 1400 h the amount of time animals spent feeding began to increase once more, reaching a peak in the late afternoon.

During the dry season, intensive feeding began later and ended earlier in both study areas; in the north there was still a decrease in feeding at midday, although it was less marked than in the wet season. By contrast, feeding rates in the south slowly built up throughout the morning to reach a peak between 1000 h and 1100 h and then decreased in the afternoon at almost the same speed.

b) Resting

Since animals in both areas spent most of their time resting or feeding, daily patterns of resting dovetailed closely with those of feeding. In the wet season, early morning counts of time spent resting were low in both areas, and then increased to a midday resting peak; as the afternoon feeding bout began, time spent resting once more decreased. The northern groups spent a mean 50% of total time each day resting in the wet season, and the southern groups a mean 56.8%.

In the dry season in both areas, early morning resting peaks were followed by a decline in time spent resting. In the south this reached its lowest point between 0900 h and 1100 h, after which there was a continuous build-up until by 1700 h resting was virtually the only activity. The northern groups deviated from this resting pattern as they did from the southern groups' feeding pattern; the slight, midday decrease in feeding the dry season was associated with a slight, midday increase in resting. The northern groups spent a mean 61.5% of total time each day resting in the dry season, and the southern groups a mean of 69.5%.

c) Range of variation

The range of variation in the amount of time spent feeding/resting in any given hour was very wide; an animal's activity at a given hour could never be predicted with certainty. Although the actual sequence and duration of activities seemed to remain fairly constant, the absolute time of day at which an activity took place apparently depended largely upon the time at which the whole sequence of activities for the day began.

d) Seasonal variation in exposure to sunshine

There was little difference between study areas in the amount of time animals spent in the shade. In both areas more time was spent in the shade during the wet season than during the dry season (Fig. 31).

Data for the first hour of recording, around sunrise, show that a considerable proportion of time was spent in the shade. The greatest

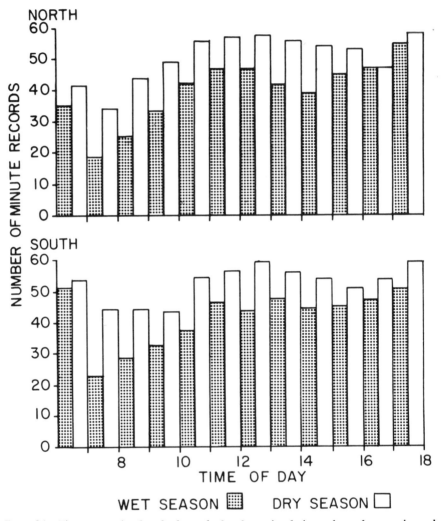

FIG. 31. Time spent in the shade each day by animals in each study area, in each
season.

exposure to direct sunlight occurred between 0700 h and 0800 h in both seasons. This was also when most sunning behavior was observed. From 0800 h, time spent in the sun decreased consistently until mid-afternoon when, in both seasons and in both areas, there was a slight reversal in this trend. By 1800 h, animals were in the shade almost 100% of the time.

5
Social Behaviors

This chapter describes interactions both within the group and be-tween animals belonging to neighboring groups. It is based on a con-tinuous record of social interactions involving the focal animal of each day's observations. Not all members of Group II could be con-sistently recognized, so it was excluded from the analysis of within-group interactions where individual recognition was important. In order to make the frequencies of interaction totally comparable among different animals, data on a particular animal were not used in the analysis if they were gathered during observations in excess of the number of hours of observation made on the least-observed animal in each season. Since the number of observation hours for each group each month was equal, and since observations were close to evenly distributed among different members of each group in each season, data lost through this process of equalization were minimal.

1. Interactions with Individuals within the Group

a) Agonistic behavior
  i) Characteristics and context
  An animal that displaced, threatened, bit, or cuffed another animal in its own group was considered to be the aggressor in an agonistic encounter. Cuffs, given with the hand, and bites, generally adminis-tered on the back of the neck or limb extremities, were both usually accompanied by a "cough" or "hack" vocalization. Animal A was said

to have displaced Animal B if, as it approached, Animal B glanced at it rapidly and leaped off at once, or if Animal B adopted a submissive posture (see below) and subsequently leaped off. Staring, or lunging in the direction of an animal were considered as threats. They resulted in the threatening animal's displacing the recipient of the threat, or in the recipient's adopting a submissive posture.

Submissive gestures occurred in response to overt agression. They included baring the teeth, with the lips drawn back tightly, rolling up the tail between the hindlegs, and hunching the back (Plate XIII). They were usually accompanied by the "spat" vocalization described by Jolly (1966) as ". . . a series of high squeaks, given in quick succession. . . . " At low levels of intensity these vocalizations were unvoiced and had the quality of a cat's purr. After giving these signals of submission, the victim might remain in its original position or leap off, depending on the persistence of the aggressor and the nature of the previous relationship between them.

In all groups, most agonistic encounters occurred in a feeding situation (see Table 5.15). Typically, the aggressor was feeding and the re-

TABLE 5.15

FREQUENCY OF AGONISTIC ENCOUNTERS AND CONTEXTS IN WHICH THEY WERE OBSERVED

Group	Total Number of Agonistic Encounters	Number per Animal Hour	Situations in which aggression occurred		
			Access to Food	Access to Resting Site	Other
I	107	0.29	98	9	—
II	187	0.42	153	10	24
III	109	0.25	84	11	14
IV	191	0.44	128	23	40

cipient came too close, or the aggressor supplanted the recipient at the latter's feeding station. Obvious spatial displacements without reference to a food source were rare and appeared to occur only at the onset of resting periods, when there was competition for what were presumably preferred resting places.

Agonistic behavior occurred less frequently in five other situations: 1) The adult female in Group II consistently rebuffed the ap-

PLATE XIII. Polo gives submissive gestures to Fred

proaches of other group members after the birth of her infant.

2) The adult female in Group II pushed away the juvenile male when, on three occasions following the disappearance of her infant, he tried to suckle.

3) Four agonistic encounters were observed that involved three animals and in which instances of apparent "redirected aggression" were seen (Richard and Heimbuch 1975).

4) Adult male Fred[1] (F) in Group III periodically grasped A♂ Polo (P) around the neck and forced the latter to groom him.

5) An increase in agonistic behavior in many contexts was seen during and prior to the mating season (see also Richard 1974).

ii) Regional and seasonal variation in patterns of interaction

The frequency of agonistic behavior varied considerably between Groups II, III and IV, but not consistently between areas (Tables 5.15 and 5.16).

TABLE 5.16

FREQUENCY OF AGGRESSION BETWEEN MEMBERS OF GROUPS II, III, AND IV (FIGURES IN BRACKETS REPRESENT NUMBER OF AGONISTIC ENCOUNTERS RECORDED DURING THE DRY SEASON, THOSE WITHOUT BRACKETS REPRESENT THOSE RECORDED DURING THE WET SEASON)

	Group II				
Recipient	♀	♂	Y♂	J	Total
Aggressor					
♀		13 (13)	34 (30)	26 (17)	73 (60)
♂			11 (4)	5 (4)	16 (8)
Y♂				10 (16)	10 (16)
J			4 (0)		4 (0)
TOTAL		13 (13)	49 (34)	41 (37)	103 (84)

(Cont.)

TABLE 5.16: Cont.

Group III

Recipient	♀NFD	♀FD	♂F	♂P	Y♂	J	Total
Aggressor							
♀NFD		(1)	4	4 (1)	9 (1)	7 (2)	24 (5)
♀FD	1 (2)		3 (5)	1 (3)	2	6 (3)	13 (13)
♂F				17 (5)	9 (1)	12 (1)	38 (7)
♂P					1	1 (2)	2 (2)
Y♂				1		2	3
J					2		2
TOTAL	1 (2)	(1)	7 (5)	23 (9)	23 (2)	28 (8)	82 (27)

Group IV

Recipient	♀FI	A♂R	♀FNI	SA ♂Q	OINT	Inf	Total
Aggressor							
♀FI		46	9 (3)	24 (20)	(1)	6	85 (24)
♂R	5*		27	17			49
♀FNI				10 (6)	(3)	2	12 (9)
SA ♂Q	1*	2*	3*				6
♂INT				(6)			(6)
Inf							
TOTAL	6	48	39 (3)	51 (32)	(4)	8	152 (39)

*These encounters were recorded March 1971.

In both study areas, agonistic encounters occurred more frequently in the wet season than in the dry season (Mann-Whitney U Test, U=1, p<.01 — Group II; U=0, p<.001 — Groups III and IV). In the south, the wet season coincided closely with the mating season but in the north, where the wet season was longer, observations were made in the wet season well before the onset of the mating season. The increase cannot, thus, be related to mating season disruption alone.

While no animals were less aggressive during the wet season, some became more aggressive than others (Table 5.17). This individual varia-

TABLE 5.17

NUMBER OF AGONISTIC ENCOUNTERS INITIATED BY EACH ANIMAL IN THE WET AND DRY SEASONS, AND INDEX OF SEASONAL CHANGE

Group	Initiator	Number of Agonistic Encounters Wet Season		Dry Season		Index of Seasonal Change
		N	%	N	%	
II	♀	73	71	60	83	1.2
	♂	16	15	8	11	2.0
	Y♂	10	10	4	6	2.5
	J	4	4	—	—	—
III	♀FD	13	17	13	48	1.0
	♀NFD	24	29	5	19	4.8
	♂F	38	46	7	26	5.4
	♂P	2	2	2	7	1.0
	Y♂	3	4	—	—	—
	J	2	2	—	—	—
IV	♀FI	85	56	24	62	3.5
	♀FNI	12	8	9	23	1.3
	♂R/INT	49	32	6	15	8.2
	SA♂Q	6	4	—	—	—
	Inf.	—	—	Not present		—

tion was not related to an animal's age and sex class; the index of seasonal change for each animal demonstrates this more clearly by giving the ratio of aggressive incidents it initiated in the dry season to those it initiated in the wet season.

b) Nonagonistic behavior

i) Characteristics and context.

Allo-grooming was the most commonly observed nonagonistic

interaction. Grooming was sometimes initiated by the groomer, but on other occasions the prospective recipient held out, or "presented," an arm towards the prospective groomer. The limbs were the focus of grooming in only 4% of bouts, however, and attention more commonly centered on the head, face, and back of the recipient — all areas inaccessible to self-grooming.

All grooming was done by licking with the tongue and scraping the tooth comb over the fur. When grooming the head, the groomer frequently clamped his hand around the recipient's muzzle (Plate XIV). Seventy

PLATE XIV. Two adults engaged in a grooming bout

percent of grooming episodes were unidirectional: one animal groomed another. In 4% of episodes, grooming was reciprocal; the two animals involved groomed each other alternately. Reciprocal and simultaneous grooming occurred in 26% of episodes; the animals sat in physical contact, each grooming the other's shoulder or back.

Play was almost always initiated by subadults and juveniles, although adults usually responded to playful approaches. Play was characterized by a relaxed, open-mouth "play-face" and involved chases and wrestling. In 73% of bouts, participants played on or within 2 m of the ground. Bouts more than 2 m above the ground generally involved two or three animals wrestling together, often hanging by their arm(s) or leg(s) alone.

Nose-touching was observed, described by Jolly (1966) as "greeting behavior"; it constituted only 5.7% of all nonagonistic interactions. The participants approached each other, touched noses briefly, and moved apart again.

Most grooming occurred during rest periods. Play behavior was seen mainly at the beginning and end of feeding and rest periods. Animals touched noses when reunited after prolonged dispersal of the group during feeding bouts.

ii) Regional and seasonal variation in patterns of interaction.

The frequency of nonagonistic behavior varied among Groups II, III, and IV, but not consistently between areas (Tables 5.18 and 5.19).

TABLE 5.18

FREQUENCY AND PATTERNS OF GROOMING IN GROUPS II, III, AND IV (FIGURES IN BRACKETS REPRESENT GROOMING FREQUENCIES IN THE DRY SEASON; THOSE WITHOUT BRACKETS REPRESENT FREQUENCIES IN THE WET SEASON)

	Group II				
Groomed One	♀	♂	Y♂	J	Total
Groomer					
♀		2 (1)	1 (1)	18 (1)	21 (3)
♂	3 (16)		2 (2)	14 (2)	19 (20)
Y♂	3 (12)	4		12 (12)	19 (14)
J	10 (7)	11 (4)	8		29 (11)
TOTAL	16 (35)	17 (5)	11 (3)	44 (5)	88 (48)

Cont.

TABLE 5.18: Cont.

Group III

Groomed One Groomer	♀FD	♀NFD	♂F	♂P	Y♂	J	Total
♀FD		2 (5)		(1)	2 (1)	6 (3)	10 (10)
♀NFD	2 (3)		2 (1)	1		1	6 (4)
♂F	2 (1)	(1)		(1)	1	3 (5)	6 (8)
♂P		(1)	4 (9)		1	2 (1)	7 (11)
Y♂	2 (1)	1	2 (1)			4 (1)	9 (3)
J	2	2	2 (5)	(1)	1 (1)		7 (7)
TOTAL	8 (5)	5 (7)	10 (16)	1 (3)	5 (2)	16 (10)	45 (43)

Group IV

Groomed One Groomer	♀FI	♂R/INT	♀FNI	SA♂Q	Total
♀FI			(3)		(3)
♂R/INT	2		1	2	5
♀FNI	10 (5)	1			11 (5)
SA♂Q	(6)	3 (1)	(3)		3 (10)
TOTAL	12 (11)	4 (1)	1 (6)	2	19 (18)

TABLE 5.19

FREQUENCY AND PATTERNS OF PLAY AND NOSE-TOUCHING IN GROUPS II, III, AND IV (FIGURES IN BRACKETS REPRESENT FREQUENCY OF NOSE-TOUCHING; THOSE WITHOUT BRACKETS, FREQUENCY OF PLAY)

Group II

Initiator Partner	♀	♂	Y♂	J	Total
♀			1 (1)	1 (1)	2 (2)
♂			3	3	6
Y♂		3		17	20
J	7	7	25 (1)		39 (1)
TOTAL	7	10	29 (2)	21 (1)	67 (3)

Group III

Initiator Partner	♀FD	♀NFD	♂F	♂P	♂Y	J	Total
♀FD		1	(2)			(1)	1 (3)
♀NFD	2			(1)	1	(2)	3 (3)
♂F					1		1
♂P		(1)			3	1	4 (1)
♂Y	1			1		12 (1)	14 (1)
J			3 (2)	1 (2)	11		15 (4)
TOTAL	3	1 (1)	3 (4)	2 (3)	16	13 (4)	38 (12)

TABLE 5.19: Cont.

Group IV

Initiator	♀FI	♂R/INT	♀FNI	SA♂Q	Inf	Total
Partner						
♀FI		(1)				(1)
♂R/INT			2 (1)	1	1	4 (1)
♀FNI	(1)	(4)		(1)	1	1 (6)
SA♂Q						
Inf				1		1
TOTAL	(1)	(5)	2 (1)	2 (1)	2	6 (8)

Only in Group II was there seasonal variation in the frequency of grooming; 88 out of a total of 136 bouts occurred in the wet season. Table 5.20 shows that the higher grooming levels of the wet season

TABLE 5.20

CONTRIBUTION OF EACH ANIMAL TO TOTAL FREQUENCY OF GROOMING IN GROUP II IN WET AND DRY SEASONS, AND INDEX OF INCREASED GROOMING

Groomer	*Wet Season*		*Dry Season*		*Index of Seasonal Change*
	%	N	%	N	
♀	24	21	6	3	7.0
♂	21.5	19	42	20	0.95
Y♂	21.5	19	29	14	1.35
J	33	29	23	11	2.63
Groomed One					
♀	18	16	73	35	0.46
♂	19	17	10.5	5	3.40
Y♂	13	11	6	3	3.67
J	50	44	10.5	5	8.80

were not due to increased grooming activity by all animals in the group; over half of all grooming at that time was initiated by the adult female and juvenile, and they were responsible for 90% of the increase over dry season frequencies. The index of seasonal change, the ratio between the number of episodes observed for each animal in each season, shows that the adult male was the only animal who initiated grooming less frequently in the wet season than in the dry. In the dry season, the adult female and juvenile groomed less than the other two members of the group, although the reverse held for the adult female with respect to being groomed; 73% of all grooming was directed at her in the dry season, as compared with 18% in the wet.

Play behavior was never seen during the dry season in either study area, but was common in the wet season. Nose-touching, in contrast, occurred occasionally throughout the year. Table 5.19 shows the direction and frequency of play behavior and nose-touching in Groups II, III, and IV. Nose-touching occurred too rarely for a characteristic pattern to emerge; all animals participated at least once, however, in this behavior. More can be said concerning play behavior, a pastime primarily for juveniles and subadults; in Group II, the juvenile and subadult both initiated bouts, and were "play-approached" more than three times as often as the two adult members of the group. More than half the play bouts recorded were between these two immature animals, and all bouts involved at least one of them. Similarly, more than half the 38 play bouts seen in Group III were between the two immature animals in that group, although 7% of bouts (3) were between the two adult females. Play was observed only six times in Group IV, although the group contained one subadult animal and one infant. Two of these bouts involved adults only: adult male Rip (R) and adult female Fanny (FNI).

c) *The mating season*

Mating is highly seasonal in *P. verreauxi*, taking place between January and March throughout the species' range. Copulation was thus witnessed only in the southern groups, in March 1971; it was said to have occurred when ejaculation took place (mounting and intromission with no ejaculation are considered separately). Copulation by 3 males with the 2 females in Group IV was observed, and on only two other occasions were males observed with erections: once during an intergroup encounter, and once during a play bout between a subadult and a juvenile.

A number of behavioral changes were recorded during the six weeks preceding copulation, or the "precopulatory period." The onset of

TABLE 5.21

SUMMARY OF EVENTS DURING THE MATING SEASON

Event	Precopulatory Period			Copulatory Period	Postcopulatory Period
	Jan. 24-31	Feb. 1-13	Feb. 20-28	March 3-6	March 7-15
Vulval flush	Present in ♀FNI (IV) on 26 and 27				
Endorsing	High frequency in ♂R(IV)			High frequency in all adult males studied	
Sniff-approach and mark sequences	High frequency in ♂R(IV)			High frequency in all adult males studied	
Roaming	1) ♂INT approaches Group IV twice 2) ♂Q(IV) leaves Group IV 3 times 3) ♂R(IV) leaves Group IV 3 times (once to approach Group III)	1) ♂INT approaches group IV twice 2) ♂LCE approaches Group IV twice 3) ♂Q(IV) leaves Group IV once.	1) ♂♂R (IV) and Q(IV) leave Group IV and make a foray into home-range of Group III 2) ♂P(III) leaves Group III twice	1) SA♂Y(III) approaches and follows Group IV 2) ♂P(III) approaches & follows Group IV; mates 3) ♂INT approaches & follows Group IV; mates 4) ♂LCE approaches & follows Group IV; is chased away	1) ♂P(III) rejoins Group III 2) ♂R(IV) approaches Group III; is chased away 3) ♂LCE approaches Group IV; is chased away

TABLE 5.21: Cont.

Event	Precopulatory Period			Copulatory Period	Postcopulatory Period
	Jan. 24-31	*Feb. 1-13*	*Feb. 20-28*	*March 3-6*	*March 7-15*
				5) ♂R(IV) ousted from Group IV by ♂P(III) and ♂ INT in turn	
Intragroup agonistic encounters	High frequency in both groups				
Intergroup encounters	High frequency in both groups			[♂F (III) mates with ♀FI (IV) during intergroup encounter on March 3rd]	
Copulation				(In order of occurrence) ♂F (III) with ♀FI (IV) ♂P (III) with ♀FI (IV) ♂ INT with ♀FNI (IV)	

these changes was marked by a two-day flushing around the vulva of
one of the females in Group IV. The precopulatory period and days
when copulation occurred are together referred to as the mating season.

The behavioral changes involved an increase in four activities, which
occurred at low frequencies throughout the year and built up during
the precopulatory period to reach a peak during the three days when
copulation took place. Two of these activities were performed by adult
males only, namely, scent-marking and "roaming" (2, b ii below).
The other two were common to all animals, namely, an increase in the
frequency of agonistic encounters both within the group and between
groups. An overall summary of these changes is given in Table 5.21,
and a fuller description is to be found in Richard (1974).

Adult and subadult animals scent-marked in a number of contexts
throughout the year. While females marked either by rubbing the anogenital region or by urinating on a trunk or branch, marking by males
included rubbing a branch or trunk with the scent gland on the ventral
surface of the throat (throat-marking), then with the tip of the penis,
usually urinating slightly in the process, and finally with the perineal
area. Although a male might perform any one part of this sequence
when marking, it was commonly carried out in its entirety. The term
endorsing was used when a male marked a spot five minutes or less
after it was vacated by a female. *Sniff-approach and mark* described
another form of scent-marking performed only by males; in this, the
male approached a female and marked the tree trunk just below her
tail (Plate XV). First he climbed the trunk below her and touched her
anus with his nose. Then he throat-marked and finally marked with
his anogenital area. This sequence was often incomplete, interrupted
by a bite or cuff from the female; in such cases, endorsing usually
followed when the female moved off.

Both endorsing and sniff-approach and marking increased significantly
in Group IV during the mating season, although no change was seen in
Group III. Table 5.22 shows changes in the frequency of endorsing in

TABLE 5.22

CHANGES IN FREQUENCY OF ENDORSING IN GROUPS III AND IV (FIGURES EX-
PRESSED AS A FREQUENCY PER ADULT MALE HOUR; ONLY DATA RECORDED
WHEN AN ADULT MALE WAS THE DAY'S SUBJECT WERE USED IN CALCULATIONS)

| Group | Jan. 1971 Vulval Change | | Feb. 1971 | March 1971 | April 1971 |
	Pre-	Post-			
III	1.95	—	2.35	1.08	0.67
IV	1.90	6.75	6.26	7.00	1.04

PLATE XV. An adult male "sniff-approach and marks" beneath an adult female

Groups III and IV between January and April 1971. Sniff-approach and mark behavior was recorded only six times in 216 hours of quantitative observation of Group IV outside the mating season. In contrast, the sequence (in complete or incomplete form) was seen 65 times in 216 hours between January 24 and March 15, 1971.

The higher frequency of agonistic behavior during the mating season (which coincided with the wet season in the south) has already been documented (Table 5.17). A further change in the nature of agonistic behavior in Group IV also took place; outside the mating season, a hierarchy could be defined on the basis of agonistic encounters over access to feeding and resting stations. Initiation of aggression and displacements were found to occur consistently in one direction. During the mating season a breakdown in this structuring occurred in Group IV; although subadult male Q took little part in events during the period of copulation, it was the only time when he was seen initiating aggression against other members of the group. No change occurred in Group III.

d) Group profiles

Descriptions of the salient features of individuals in Groups II, III, and IV are based on data given in more general form in the preceding sections and in Tables 16-20. An analysis of the effects of age/sex class differences and season on the pattern of dispersion in Group III is also presented here. This group was chosen for the analysis because of its relatively large size and balanced age/sex class composition. As a measure of group dispersion, repeated observations on the distance between the focal animal and its nearest neighbor were made (see also 5, 5,3 above). Results were expressed as a percentage of the total number of observations made on a class of focal animals in a given month. For example, if there were 50 observations on young males during December and in five of those the nearest neighbor, an adult male, was in distance-category one, then the wet-season value for Young Male (focal), with Adult Male (neighbor) in distance category one would be 0.10.

These data were analyzed using the techniques and assumptions of Multivariate Analysis of Variance and Multivariate Linear Regression. Both methods can be expressed in the general linear model:

$$Y = X \quad B + E$$
$$(n \times p) \quad (n \times q)(q \times p) \quad (n \times p)$$

where: n=number of observations (120 in this study), p=number of dependent variables (five), and q=number of coefficients (32). Data in the five distance categories were used as a five-dimensional dependent vector, each distance-category comprising one dimension. Dispersion patterns were modeled as being dependent on three factors ("treatment effects" in MANOVA) and their interactions ("interaction effects"). Treatment effects were the focal individual (Focal), age/sex class of the nearest neighbor (Neighbor), and the season.

Interaction effects allowed for the possibility of more idiosyncratic behavior by assuming that combinations of the treatment effects may have had distinctive results. The interaction effects were: Focal and Neighbor, Focal and Season, Neighbor and Season, and Focal and Neighbor and Season; for example, if adult males spent a disproportionate amount of time grooming other adult males, irrespective of season, this would be reflected as a large positive value in distance-category one for the Adult Male/Adult Male (Focal/Neighbor) interaction effect.

In the X matrix, each row identified which treatment and interaction effects may have influenced the observations recorded in the same row of the Y matrix. The X matrix was designed so that the sum of any set of estimated treatment and interaction effects would equal zero and reflect average deviations from the grand mean. With the X matrix constructed, the B matrix was estimated using least squares criterion:

$$B = (X'X)^{-1} X'Y$$

In order to determine whether an estimated value in the B matrix was significantly different from zero, Bayesian distribution theory was used to generate credibility intervals, and those with greater than 95% posterior probability of being different from zero were used as the basis for statements concerning group dispersion. (A fuller discussion of this use of MANOVA techniques and of the problems and assumptions involved is given in Heimbuch and Richard, in prep.)

Group II: no agression was directed at the single adult female, who was herself the most frequently aggressive member of the group (at one stage, this aggressiveness was probably related to protecting her newborn infant). She groomed least and received most grooming; she and the juvenile were each groomed more than twice as much as either the adult or subadult males. What little grooming the female did initiate was generally directed at the juvenile. Relationships appeared least structured between the two immature animals, who played together extensively during the wet season; each was observed initiating aggression against the other, a bidirectionality found uniquely in this dyad.

Group III: no aggression was directed at the two adult females in this group, either. Occasional hostile interactions between the two of them were prolonged and of unpredictable outcome. Although submissive to these females, adult male Fred directed aggression at other group members more commonly than did either of them. He was in fact the most frequently aggressive animal in the group, over half his interactions being with the other adult male, Polo. In this group, aggression was bidirectional, not only between the two adult females, and between two immature animals, but also between adult male Polo and the subadult male.

One of the adult females, Freda (FD), and one of the adult males, Polo, groomed more often than other animals. In both cases, their grooming was directed primarily at just one other animal; nearly half Freda's grooming was directed at the juvenile, and nearly three-quarters of Polo's at Fred. Predictably, the juvenile and Fred were both groomed more than twice as much as any other animals; the least groomed were Polo and the subadult male.

Concerning the dispersion tendencies of this group, all animals tended to spend their time in distance-categories 2, 3, and 4 relative to other members of the group. Adults of both sexes tended to spend more time in distance-category 1 when their nearest neighbor was an adult of the same sex. In other words, males "preferred" each other as nearest neighbor, as did females. For the males this result must be interpreted with caution: aggression was frequent between them, so the "preference," or proximity, may have been due to a reduction of interanimal distance during agonistic encounters rather than to mutually high tolerance. Some preference was also shown for the subadult male, particularly during the dry season; when he was their nearest neighbor in that season, the adult males spent more time in distance-category 2 relative to general tendencies. The wet season — coincident in this area with the mating season — saw a general increase in interanimal distance; the two adult males spent more time in distance-category 5 relative to the subadult, and the adult females distanced themselves from each other as well as from other members of the group. There was only one anomaly; although the general wet-season distancing effect was seen among the adult males and females, adult females did spend significant amounts of time in category 1 relative to adult males, that is, in physical contact with them. Irrespective of season, the two immature members of the group spent a lot of time in close proximity to each other.

Group IV: observations were made on this group when the two adult females were in oestrus as well as when they were anoestrous. Outside the mating season, adult female Fiona (FI) was the most frequently aggressive animal in the group, and also the only one against whom aggression was never directed. Fanny, in contrast, was the only female in any group to be the object of aggression from a male: Rip initiated aggression against her and displaced her regularly. Outside the mating season, she could displace the subadult male and in fact he directed no aggression at anyone during that period. When the adult females were in oestrus, this structure was upset; the subadult male initiated aggression against all other members of the group, and Rip began to initiate aggression against Fiona.

Both adult females groomed infrequently, and when they did they almost always groomed each other. The subadult male was the only animal to groom all other members of the group often.

2. Interactions with Other Groups and Nongroup Individuals

a) Outside the mating season

  i) Characteristics and context of intergroup interactions

An interaction between two groups was said to have occurred when members of each group oriented visually and/or auditorily toward the other and when I could see or hear at least two members of each. Such a definition included interactions of a wide range of intensity, but within this range three general categories* could be distinguished that included almost all interactions. The first category contained interactions during which animals appeared to be at a low level of arousal: members of each group stared in the direction of animals in the other group (although direct visual contact was not always made), and usually one or two animals growled softly. Although these interactions interrupted ongoing activity, they did not result in a chase or battle and indeed no apparent or immediate move was made by either group in response to the other's presence. Category 1 was called *Growl, Stare, and Remain* (GSR).

The second category, *Growl, stare, and move*, (GSM) included interactions at a seemingly intermediate level of arousal: animals from each group stared, growled, and then moved gradually away from each other. As in category 1, ongoing activity was disrupted, although no

---

*N.B. Although formulated independently of Hausfater's (1972) categories of intergroup interaction among rhesus monkeys *Macaca mulatta*, it is of interest to note that they resemble his categories closely.

major confrontation occurred; category 2 was distinguished by an immediate spatial displacement by both groups, resulting in an increase in distance between them.

Categories 3 and 4 both referred to interactions occurring at a high level of arousal. Category 3 included chase/flight interactions; the interaction was recorded as a chase (CH) if the primary group under observation pursued the group it had encountered, and as flight (FL) if the primary group was itself chased. Such encounters generally appeared to involve high levels of energy expenditure; animals moved from vertical trunk to vertical trunk through the forest at a speed usually observed in response to other species, such as *H. sapiens.* Before the chase, each group usually formed a compact unit and sat growling, staring, head-jerking, scent-marking, and sometimes "sifakaing" (3:4, a) at the other. With no apparent warning, members of one group would then launch themselves after those in the other, who immediately took to their heels: the distinguishing features of this category were the intensity of the interaction and the unidirectional chase.

During category 4 interactions, animals appeared to be at an equal or even higher level of arousal; the category was characterized by reciprocal chasing or "battles" (B), well described by Jolly (1966):

> The troops often seem to mix, with animals leaping about in chaos. Each group, however, keeps its own orientation, the troops facing their goals like sets of opposing chessmen . . . everything depends on a fast, formal pattern of movement, each animal occupying sections of tree rather than opposing individuals of the other group. Each tense leap, therefore, carries an attacker toward a particular undefended area of tree, *not* into contact with an enemy. The troops, therefore, move in reciprocal formations.

Sometimes these battles ended with one group chasing the other, and sometimes they ended for no apparent reason with both groups moving off to feed in different directions.

Table 5.23 divides all but one of the observed intergroup interactions of each of the study groups into these four categories, and shows the ongoing activity and time at their onset. Predictably, almost all intergroup encounters occurred when animals were feeding or moving; since the activities of the neighboring study groups and, it is presumed, of all groups in an area were quite closely synchronized, it was during feeding bouts or when moving across their range that groups were most likely to meet.

TABLE 5.23

TOTAL NUMBER OF EACH ENCOUNTER TYPE, ACTIVITY, AND TIME AT ONSET OF ENCOUNTER FOR EACH GROUP

Group	Encounter type	Total Number*		Activity at onset**			Time at onset			
		n	%	Feed/forage	Move	Rest	0600-0900h	0900-1200h	1200-1500h	1500-1800h
I	1) GSR	2	7	—	1	1	1	1	—	—
	2) GSM	5	18	3	—	—	1	3	1	—
	3) CH/FL	9	32	4	2	3	2	4	2	1
	4) B	12	43	9	—	1	1	4	5	2
II	1) GSR	14	33	10	2	2	3	2	6	3
	2) GSM	5	12	3	—	1	4	1	—	—
	3) CH/FL	17	39	7	5	1	2	6	8	1
	4) B	7	16	2	1	—	—	3	3	1
Totals for Northern Groups		71		38	11	9	14	24	25	8
III	1) GSR	3	17	2	—	1	3	—	—	—
	2) GSM	5	28	4	1	—	4	1	—	—
	3) CH/FL	1	5	1	—	—	—	1	—	—
	4) B	9	50	5	—	—	8	—	1	—
IV	1) GSR	2	11	1	—	—	2	—	—	—
	2) GSM	4	22	2	2	—	3	1	—	—
	3) CH/FL	3	17	3	—	—	1	2	1	1
	4) B	9	50	4	—	1	8	—	1	—
Totals for Southern Groups		36		22	3	2	29	5	2	—

* For each study group this includes both encounters recorded when the focal animal was a member of that group and encounters recorded when the focal animal's group interacted with that group.

** Activity at onset is given only for the group to which the focal animal belonged, i.e., when an encounter between Groups I and II occurred and the focal animal belong to Group I, only the activity of Group I at onset could be recorded. (The activity of the focal animal is here assumed generally to reflect "group activity.")

As noted above, one interaction outside the mating season did not fit the general classification given here. It involved Group IV and a neighboring group to the south. When observations were begun in September 1970 and neither of the southern groups was habituated, Group IV contained seven adult animals and two infants; all seven were individually recognizable, and each morning all or some combination of these animals were located sunning themselves before they dispersed, fled, and hid from me. This group was still not habituated when observations were terminated at the end of the month and no quantitative data were collected on it at that time. When the group was recontacted and habituated in January 1971, it contained four adult animals (Table 4) and one infant. Thereafter, these animals habitually traveled and rested together as a group with only one exception; one morning in mid-February they were found sunning themselves with three unknown animals in the south of their range. All eight animals were sitting in a single *A. ascendens* with no more than 2 m separating one animal from another. Shortly after I found them, the three unhabituated animals dispersed rapidly and were not seen with Group IV again.

ii) Frequency and patterns of intergroup interaction

Intergroup encounters were commoner in the north than in the south (Table 23), and in both study areas they were generally more frequent in the wet season than in the dry (Fig. 32). In the north, Group I took part in 28 interactions with other groups during 432 hours of observation (0.065 per hour), and Group II in 43 (0.099 per hour). Of these, 13 were interactions between Groups I and II. Encounters occurred throughout the day, although less commonly toward evening. In the south, Groups III and IV each interacted with another group on 18 occasions (0.041). Eleven of these encounters were between Groups III and IV. In contrast with the northern groups, the southern groups tended to interact with neighboring groups only in the early morning; 80% of all encounters occurred between 0500 h and 0900 h.

There was some variation in the type, or category, of encounter that each group most commonly experienced. For example, "battles" accounted for 40-50% of Groups I, III, and IV's encounters and for only 16% of Group II's. There was no evidence that type of interaction varied consistently between regions. Indeed, apart from the frequency of interactions, the only aspect of intergroup encounters that differed markedly between regions concerned their spatial location; in the north, encounters occurred in many parts of the home-

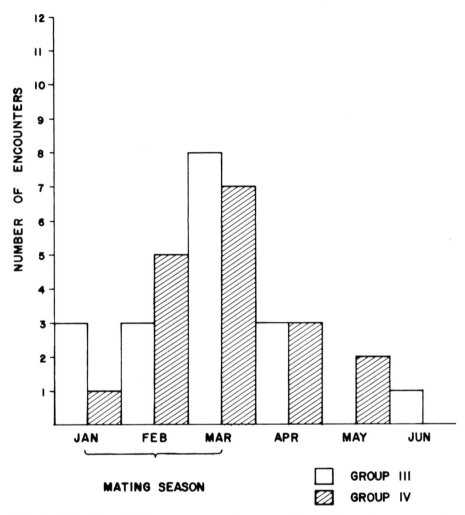

FIG. 32. Number of intergroup encounters experienced by each group, each month.

range of each group and did not appear to denote a boundary. In the south, encounters occurred around the periphery of the area of forest used exclusively by the group (Fig. 12).

No consistent pattern of interaction was found between Group I and II, but analysis of encounters between Groups III and IV indicates that a structured relationship, or some form of "intergroup dominance," did exist between them. Out of a total of eleven encounters, two were in category 2: animals from opposing groups growled at each other, stared, and then moved apart. Two consisted of "battles" terminated by the retreat of both groups into their respective ranges. The remaining seven were all "battles" that ended with Group III chasing Group IV deep into the latter's range.

iii) Interactions with nongroup individuals

Roaming was the term used to describe the behavior of animals, singly or in pairs, who detached themselves from their own group and made long forays into the home-ranges of other groups. This behavior was encountered only in adult males and was rare outside the mating season.

In the northern study area, only one interaction between a nongroup individual and one of the two study groups was witnessed, although two adult males entered Group I in the course of the study; in the two latter cases, I was not present during the period of assimilation. The single observed interaction involved an adult male and members of Group I, and occurred in August 1970. At that time, Group I contained two adult males and five adult females:

1012 h Unknown male appears on periphery of group, adult male N immediately chases him away.

1146 h Unknown male has spent last hour and a half about 30 m from group. Now tries to approach group again. Again adult male N chases him away.

1252 h Tries to approach again. Adult male N chases and catches him. The two animals fall about 6 m to ground, wrestle, and then spring apart. Adult male N returns to rest of group, who have ignored entire sequence of interactions, and sits licking bleeding wound on foot. Unknown male disappears west.

In the south no interactions between the groups and nongroup individuals were seen outside the mating season.

b) *During the mating season*

i) Frequency and patterns of intergroup interaction

During the mating season, Groups III and IV were both involved in many more encounters with neighboring groups than during other months (Fig. 32).

During the mating season, 79% (11/14) of Group III's encounters were with Group IV, and 69% (9/13) of Group IV's were with Group III. Equivalent figures outside the mating season were 25% (1/4) and 40% (2/5).* The ranges of at least two other groups overlapped with each of Groups III and IV's ranges; thus, outside the mating season the two study groups interacted equally often with all neighboring groups, but during the mating season they interacted much more frequently together than would be expected by chance. It is unlikely that my presence accounts for the discrepancy during the mating season, in that I was equally likely to inhibit the approach of unhabituated neighboring groups throughout the study.

ii) Interactions with nongroup individuals

During the mating season, evidence both from the study group males and from the arrival of unknown males in the study area showed "roaming" to be common at that time; twenty-two instances were recorded of adult males attempting to approach groups other than their own (Table 21). These excursions often culminated in fierce fights between males from different groups, but the response of animals to nongroup males did vary, apparently according to the identity of the male in question. This point is illustrated by summarizing briefly some of the events between March 3 and 6, the four days when copulation with females in Group IV was observed.

March 3: at the end of an encounter between Groups III and IV, Fiona (IV) was seen surrounded by the adult and subadult males of each group. Each male in turn attempted to approach her but she bit and cuffed at all of them. Fred (III) finally moved off and she followed him immediately. Copulation followed, with little interference from other members of the group. Fred (III) then rejoined the rest of Group III who were feeding within the limits of their own home-range. For the rest of the day, however, Fiona (IV) repelled all attempts by Rip (IV) to mount her, and she and Rip (IV) repeatedly chased away Polo (III), who spent the whole day on the periphery of Group IV trying to approach Fiona.

March 4: Polo (III) finally managed to chase a severely wounded Rip (IV) out of Group IV, and during the rest of the day Polo

*The discrepancy between these figures and those given in Richard 1974 is due to the differentiation in this analysis between encounters between groups and encounters between groups and nongroup units or individuals.

(III) copulated intermittently with Fiona (IV), who was now a willing partner.

March 5: When contact was made with Group IV in the morning, Polo (III) was not with the group. (He was not seen again until March 10, when he reappeared with Group III, his nose deeply gashed.) Rip (IV) attempted to approach the group, but was chased away by the sub-adult male Quip (IV) and was not seen again. In the afternoon an unknown male, Intruder (INT), approached the group, mounted Fanny (IV), and copulated with her repeatedly. None of the animals tried to chase him away, and Fanny (IV) did not try to escape when he mounted her.

March 6: Intruder again mounted and copulated with Fanny (IV). In the afternoon, all four members of Group IV repeatedly chased away another unknown male, Luke (Lce), who attempted to approach the group. The composition of Group IV did not change again before the end of the study.

Thus, while Fred (III) gained access to Fiona (IV) with little or no resistance from Rip (IV), Polo (III) had to fight for 24 hours and chase Rip (IV) from the group before copulating with her. When Rip (IV) attempted to reenter the group, he was driven off, but shortly thereafter another adult male, Intruder, gained entry without resistance. Finally, yet another nongroup adult male, Luke, was chased away by the whole group when he tried to approach.

The differing responses of the two adult females in Group IV to the sexual advances of these males was equally striking: on March 3 Fiona (IV) rejected all males except Fred (III). On March 4 she rejected both Polo (III) and Rip (IV) until the former drove the latter from the group; then she copulated with Polo (III). On March 5 and 6 Fanny (IV) presented to Intruder as soon as he approached her, but on March 7 she helped chase Luke away from the group.

Finally, a single interaction on March 7 between a group and nongroup animals should be noted, for it was unique, falling outside the range of interactions described so far:

1015 h   "Battle" between Groups III and IV (Group IV now composed of Fiona, Fanny, Intruder, and Quip) terminated by Group IV moving back toward center of their range.

1027 h   Fred (III) and Yip (III) follow Group IV.

1034 h   Fred and Yip approach Group IV and play with all four members.

1101 h   Animals have been chasing and wrestling for twenty minutes, intermittently showing characteristic "play-face."
1102 h   Fred and Yip leave, seemingly without harassment, and return to rest of Group III.

Notes to Chapter 5

1. Animals in Groups III and IV are named rather than identified with a letter, to facilitate reading.

6
The Meaning of
"The Group" for *P. verreauxi*

1. Group Size and Composition

P. verreauxi lives in apparently stable populations composed of small groups, each containing three to twelve animals. In this study it was found that while the sex ratio for a given population approached 1:1, in a single group there could be a striking excess of males or females. Mean group size was similar in all but one of the forests surveyed; animals lived in larger groups in tamarind-dominated gallery forest at Antserananomby, in the west of the island.

Since the sex ratio varied so much from group to group, it was possible to find groups fitting the descriptions *one-male group, family group,* and *multi-male group,* and no support was found for Petter and Peyrieras's generalization (1974):

> With the Sifakas [*P. verreauxi*], as with the *Avahi* (i.e. in all the indriidae), the groups would normally have a strictly family basis . . . . But in zones which are more-or-less degraded or transformed . . . it is possible that the family groups are modified to form larger groups. . . . it is degradation of the environment which favors the formation of larger groups.

Pollock (1975a) reports a family group structure as characteristic of *Indri indri,* but evidence gathered in this study tends to contradict

158

the assertion as far as *P. verreauxi* is concerned. First, the largest groups of *P. verreauxi* censused were at Antsernananomby, which was undoubtedly the least disturbed and most extensive of all the forests surveyed. Second, a self-replicating family group system must incorporate behavioral mechanisms leading to the dispersal of juvenile and subadult animals from the maternal group; yet Pollock (1975) points out that even for *Indri indri* there is still no "rigorous proof of a disciplined state of monogamy." Only among the Hylobatids have such mechanisms been documented (Chivers 1974). In *P. verreauxi*, no evidence of tension between subadult and adult animals was found in the form of higher frequencies of aggression or peripheralization of subadults: young males were not driven from the group by the adult male. A third factor arises in connection with reproductive behavior. The term *family group* implies that mating takes place between the two adults in the group; yet in this study adult males were observed copulating with females belonging to other groups and the adult male in Group IV did not in fact succeed in mating with either of the females in Group IV. Finally, one general point should be made with reference to Petter and Peyrieras's argument; degradation of the environment need not necessarily favor the formation of large groups. To speak simply of *degraded forest* as an ecological category is an oversimplification; the characteristics of both the original forest and the degradation that has taken place may vary. The formation of larger groups is likely, when it happens, to be a function of many factors, including predation pressures and the size, distribution, and seasonality of food resources.

2. Structuring of the Group

In the past, primate societies were commonly viewed as simple hierarchical structures (e.g., Zuckerman 1932). Animals, almost always males, were ranked according to their social dominance, variously defined according to the frequency or direction of aggression, priority of access to food, or frequency of copulation. Females were generally excluded from the analysis. Holistic theories of social dominance generated from those early studies were gradually abandoned as awareness of the complexity of primate social structure increased, but a more limited concept of dominance is still usefully being applied in a number of field studies. In these, the dominance hierarchy is not considered an *a priori* assumption but rather the structure *resulting from* directionally consistent interactions between pairs of animals in carefully specified contexts. Jolly (1966), for example, noted that the ranking order of adult male *Lemur catta* inverted when frequency of sexual behavior was substituted for frequency of aggressive behavior as the criterion of rank. In a discussion of the social structure of the yellow baboon, Hausfater (1975) commented:

> Studies that focus on hierarchy, rather than consistency of paired relationships, run the strong risk of neglecting or ignoring important information . . . . Reversals, i.e. agonistic bouts in which the presumed subordinate defeated the presumed dominant animal, are often viewed as minor perturbations of some stable underlying social structure . . . . [They] may actually be the result of fundamental biological processes, such as sexual cycling or maturation.

In *P. verreauxi* the frequency of interaction between animals was low, and most intragroup agonistic interactions occurred with reference to access to a feeding station. Within this context, the outcome of interactions between most pairs of animals could be accurately predicted and the resulting hierarchy was termed a *feeding hierarchy,* rather than a *dominance hierarchy,* in order to emphasize its limited specificity. In all four groups, the highest ranking animal in the feeding hierarchy was an adult female, but rank was not necessarily a corollary of sex: in Group IV, Rip always displaced Fanny.

There were two exceptions to the general rule that agonistic interactions in this context were unidirectional:

1) The two adult females in Group III, who were both unchallenged by other group members, occasionally engaged in protracted aggressive interactions with each other, of unpredictable outcome. This contrasted with the situation in Group IV, where Fiona always displaced Fanny.
2) Subadult and juvenile animals were always displaced by adults in agonistic encounters, but no stable relationship appeared to exist between the subadults and juveniles themselves; aggression between them was two-way.

Outside the context of food, "minor perturbations" were apparent in some of the sets of relationships collectively referred to as the feeding hierarchy. It is argued that such "perturbations" should not be dismissed as insignificant but rather that they reflect important facets of the social structure of the groups studies. Consider first the high frequency of aggression shown by the adult female in Group II during the dry season. She gave birth to an infant in the middle of the dry season, and thereafter other members of the group repeatedly tried to approach her and to groom or handle the infant. Sixty-eight percent of this female's aggression during the dry season was directed toward animals trying to gain access to her infant during the first two weeks after its birth. She subsequently tolerated their approaches and allowed them to carry as well as handle the infant. In the wet season, in contrast, her aggression can be understood within the structure of the feeding hierarchy: 67 out of 73 encounters were over access to a feeding station.

The mother's protectiveness of her infant and her and its "special status" provide one example of a set of relationships not usefully interpreted in a hierarchical sense. Two further instances come from Group III. While in Group IV the frequency of aggression was positively correlated with the direction of aggression in the feeding hierarchy, data from Group III show that this association was not inevitable. Fred was the most frequently aggressive animal in the group, although he ranked third in the feeding hierarchy. Forty-nine percent of his encounters were with Polo, and many of these occurred outside a feeding context. This frequent, and often apparently gratuitous, aggression toward Polo is open to at least two interpretations: it could be seen as an incipient attempt to drive the second male out of the group, in line with Petter and Peyrieras's (1974) hypothesis that the family group is the fundamental unit in *P. verreauxi*. It could also be attributed simply to variation among individuals. Since other adult males were observed to coexist in other groups with a high level of mutual tolerance, I would argue for the latter interpretation; the Fred/Polo relationship reflected individual idiosyncrasies rather than some general pattern of adult male interaction. I am tempted to say that these animals simply disliked each other, bearing in mind Hinde's comment (1974):

> Students of animal behavior are so aware of the horrors of anthropomorphism that they sometimes shy away from the most interesting aspects of their subject matter: the over-simple view they get could be corrected by a little disciplined indulgence.

At the very least, it underlines the importance of considering interactions between individuals, individuals with "histories" and "kin" relations, rather than between age/sex classes, although in only a few field studies to date have most, or any, animals been consistently identified or their genealogical relationships determined (e.g., Chivers

1974; Van Lawick Goodall 1968; Hausfater 1975; Simpson 1973; Struhsaker 1975).

If one turns to nonagonistic behavior, the notion of dominance must be used equally cautiously: adult females generally initiated less grooming and were more commonly groomed than other animals. This suggests a positive correlation between frequency of being groomed and rank in the feeding hierarchy — a common relationship in many primates: " . . . the majority of the allogrooming bouts in Old World primates are against the dominance slope of the hierarchy prevailing in these communities" (Sparks 1967). However, frequent grooming as a function both of maternity and of harassment of one animal by another complicated grooming relationships and removed the simple linearity of the feeding hierarchy; over three-quarters of the grooming initiated by the adult female in Group II was directed at the juvenile in that group, who was, presumably, her offspring. Similarly, half the grooming initiated by Freda in Group III was directed at the juvenile. The intense relationship between the adult males in Group III was also reflected in grooming frequencies; rather than grooming predominantly the two high-ranking adult females in the group, Polo directed most of his grooming at Fred. Often it was the latter who forced him to do so, by grasping him around the neck and forcing his head into his, Fred's, fur.

The only actual reversal of the feeding hierarchy occurred in Group IV during the mating season. Although Quip took little part in events during the days of copulation, he did initiate aggression against other members of the group so that agonistic behavior became, briefly and uniquely, bidirectional. In Group III, in contrast, no reversal was seen in the hierarchy, nor was there any increase in aggression by the "subordinate" adult male in the group, Polo. However, taking events in Groups III and IV together, it was clear that neither access to females nor frequency of copulation was necessarily the prerogative of males who ranked high in the feeding hierarchy prior to the mating season. Indeed, the mating season appeared to act as a catalyst, producing enduring changes in rank of animals in the hierarchy.

Females were receptive only to males who asserted themselves in the immediate situation, but the basis for such assertiveness seemed to vary; in the case of Fred, his unchallenged access to Group IV's adult female, Fiona, may have rested on recognition of his unassailable rank by other animals in both Groups III and IV. Polo, in contrast, fought and drove out the resident male in Group IV and only then was Fiona prepared to copulate with him. He in turn was driven out by another male, Intruder. Polo then returned to Group III, but Intruder remained in Group IV.

Events of the mating season appeared to engender social chaos, but it is argued that an underlying pattern — if poorly substantiated as yet — was present. A male who ranked high prior to the mating season and retained this status during it mated with females in one or more other groups but remained a member of the group to which he previously belonged (e.g., Fred in Group III.) A low-ranking male prior to the mating season who fought his way to higher status in his own or another group by ousting the originally high-ranking male, may stay in that group, retaining his high status after the mating season (presumptively Intruder, although his previous status was not known). These are the success stories. In contrast, a male who ranked high prior to the mating season may be challenged and displaced during it (Rip, in Group IV). Similarly, a low-ranking male prior to the mating season who fought his way to higher status may in turn be ousted and forced to revert to his previous low status (Polo, in Group III). Under this model (see also Table 6.24), Rip recognized Fred as a high-ranking neighbor and hence as no threat to his own position; in contrast, Polo, a low-ranking male prior to the mating season, did constitute a challenge to Rip's position and fighting resulted.

3. The Group as Part of a Wider Network

In 1962 Altmann defined a society, or social group, as consisting of conspecific, intercommunicating individuals bounded by frontiers of far less frequent communication. In 1969 Struhsaker discussed the problem of operationally defining social groups of forest-living monkeys, and listed a polythetic set of distinguishing characteristics: members of a social group have the majority (at least 80%) of their non-aggressive social interactions within this social network; group members remain together for long periods and occupy the same home-range; they have distinct social roles within the group, and exhibit different behavior toward nongroup conspecifics; social groups are relatively closed to new members. Both definitions tacitly recognize that " . . . most classificatory systems, if pressed far enough do not work" (Hinde 1974), but at the same time both emphasize the importance of the "social group" as a more or less closed system. This emphasis has also been apparent in many field studies; the theoretical importance of some wider unit, or population, as the ultimate reproductive unit or gene pool is acknowledged, but the practical focus of research is ex-

TABLE 6.24

POSSIBLE CHANGES IN ADULT-MALE STATUS AS A RESULT OF THE MATING SEASON

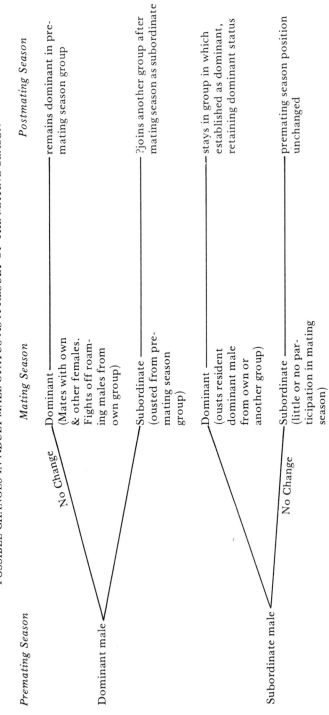

Premating Season

Dominant male

Subordinate male

Mating Season

Dominant
(Mates with own
& other females.
Fights off roam-
ing males from
own group)

Subordinate
(ousted from pre-
mating season
group)

Dominant
(ousts resident
dominant male
from own or
another group)

Subordinate
(little or no par-
ticipation in mating
season)

No Change

No Change

Postmating Season

remains dominant in pre-
mating season group

?joins another group after
mating season as subordinate

stays in group in which
established as dominant,
retaining dominant status

premating season position
unchanged

clusively on the social and sexual behavior of a single "social group."

Today, an increasing number of studies indicate that some of these "social groups" are more "open" than had been thought, suggesting that future studies of "social behavior" should include more detailed analyses of, within- *and* between-group relationships. For example, in a description of the social organization of the common white-handed gibbon in 1968, Ellefson commented, "Social behavior and social communication in a gibbon species can be defined only if at least two (preferably, more) groups living adjacent to one another are under observation . . . a social group (a communicating society) varies in composition depending on the pattern of communication in question . . . " A study of baboons living on forest fringes in Uganda showed that adult and large juvenile males and adolescent females changed groups periodically (although adult females and small juveniles did not) (Rowell 1969). Sade (1972) found that most adult male rhesus monkeys on Cayo Santiago left their natal groups during adolescence. Fully adult rhesus in Northern India have been observed changing groups during the mating season and it appears that "annual shifting of males between groups is basic to the species" (Lindburg 1969). Even adult female mobility has now been documented (McGinnis 1973, cited in Hinde 1974; Nishida 1968), although its incidence appears to be rarer than that of male mobility among the primates as an Order.

The groups observed in this study possessed some but not all of Struhsaker's (1969) polythetic set of features distinguishing a "social group," and Ellefson's comments apply as well to *P. verreauxi* as to the gibbon. What is important, though, is not the nomenclature but the quality and intensity of the animals' relationships. It is my hypothesis that the four groups functioned primarily as foraging groups within a broader social and reproductive context; clusters of foraging groups together composed larger intergrading units, or neighborhoods. Social ties within the foraging group were maintained through persistent spatial proximity and intermittent agonistic and nonagonistic interactions. The social network of the neighborhood was maintained through encounters of varying intensity and outcome between foraging groups, the movement of animals between foraging groups, and copulation between members of different foraging groups during the mating season. Jolly (1966) had a similar view: " . . . *Propithecus* or *Lemur* social behavior does not stop with the interactions within the families. . . . " It was she who introduced the concept of the *neighborhood*, noting that their world was " . . . not confined to the other members of the same family but extended through repeated acquaintance to nearby animals."

Many aspects of the data support this idea. First, although there appeared to be an upper limit on the size of foraging groups, there was not the norm of group composition, or "central grouping tendency" (Carpenter 1953) that characterizes many primate groups (e.g., Chivers 1974; Kummer and Kurt 1963). It is reasonable to assume that ecological parameters place an upper limit on the size of groups-as-foraging-units and that the normative adult sex ratio for groups-as-reproductive-units is adjusted for a given population to optimize survival and reproductive success. In *P. verreauxi*, the absence of any such optimal ratio suggests that the foraging group's survival is not contingent upon its sex ratio and that this group is itself of little or no importance as an integral reproductive unit. Second, foraging groups were by no means "closed" systems; two males disappeared from one of the northern groups and two new adult males replaced them in the course of the study. Furthermore, the mating season in the south was a period of extensive "reshuffling" of males between groups.

Events of the mating season also underlined the strong across-group as well as within-group relationships. Groups III and IV interacted with each other much more frequently during that period than with their other respective neighbors, and Group III spent a disproportionate amount of time in the narrow overlap area between their ranges. It is unlikely that this uniquely high frequency of interaction occurred because other, unhabituated groups fled from either Group III or IV when they saw me; outside the mating season, the study groups interacted equally often with all their neighbors. Throughout the study, the relationship between the two study groups appeared to be structured: interactions commonly began as confrontations but ended with Group III chasing Group IV. In other words, there seemed to be some form of intergroup dominance.

Although the underlying cause of the unequal relationship between Groups III and IV is hard to assess, I believe that it involved across-group recognition of individual animals and, particularly, of males. When adult males Fred and Polo from Group III attempted to approach the oestrous females in Group IV, Rip responded differently to them. He watched the high-ranking Fred copulate with Fiona without interfering, but he fought fiercely to prevent the low-ranking Polo from entering the group. Similarly, Fiona rejected the advances of all males except Fred. (Copulation, it should be noted, was seen only between animals belonging to different groups.) The sample is small and further observations are clearly needed, but the evidence does point to individual recognition of identity and status among animals belonging to different groups. These individual relationships could be due to animals' spending their infancy and adolescence in the same group, as kin or unrelated peers, or simply to a familiarity born of many years of interaction in and out of the mating season when their respective groups met. The one instance of play behavior between Group IV and two members of Group III indicates that animals from different groups could, occasionally, interact in a relaxed and friendly fashion. This in itself bespeaks prolonged and close familiarity. The hypothesis presented here, therefore, is that *interindividual* relationships regulated *intergroup* relationships: Rip (IV) recognized Fred (III) as a higher-ranking male during the mating season and, outside the mating season, Group IV generally gave way to Group III.

Finally, some consideration should be given to the functional significance of this system of small but closely interconnected groups or foraging parties. If ecological factors select for small-sized foraging groups, then the evolution of a series of larger, interconnecting reproductive units might be an important counter to inbreeding. It has recently been argued, however, that " . . . selection is both the primary cohesive and disruptive force in evolution, and that the selective regime itself determines what influence gene flow (or isolation) will have" (Ehrlich and Raven 1969). It cannot, thus, be confidently postulated that the network described here is a mechanism to promote gene flow; gene flow aside, it seems unlikely that the actual system of mating, with its fierce inter-male fights, evolved simply to counter inbreeding. These fights probably led to intrasexual selection between males. Access to females and, by inference, biological paternity, appeared to depend upon the fighting ability, strength, and endurance of the adult male. This may have been proved in the previous year or years, or may only have been manifested in the current mating season. Fighting ability, strength, and endurance cannot necessarily be equated with overall fitness but must be some measure of an animal's capacity to survive prolonged periods of stress.

7
An Overview of Behavioral Variability in *P. verreauxi*

1. Maintenance Behaviors

a) Ranging and group dispersion

At least three contrasting patterns of group dispersion exist in *P. verreauxi*. In the northern study area the home-ranges of both groups studied overlapped extensively with those of other groups, and intergroup encounters occurred throughout these areas of overlap. The resident group made intensive use of pockets of forest scattered throughout its range. In the south, home-range size was similar to that in the north, but overlap between groups' ranges was minimal and intergroup encounters took place only in the restricted areas of overlap. The resident group made intensive use of a large, central block of forest in its range. Finally, at Berenty, *P. verreauxi* live in home-ranges less than half the size of those described in this study (Jolly 1966). The percentage overlap between the ranges of neighboring groups at Berenty was almost identical to that found in the southern study area, and Jolly described each group as having a "nucleus of territory that others did not penetrate." She suggested that " . . . the territory and range of a group represent some minimum combination of behavioral-ecological requirements." Petter (1962a) likewise commented that *P. verreauxi* territories varied in size, being large where the forest was sparse and small where it was thick.

The concept of *territoriality* has been variously defined and applied

168

to many primate species, as well as to the birds and mammals for which the concept was originally developed (for reviews, see Bates 1970; Noble 1939). Two different emphases have been placed on the term; on the one hand, *territory* has been defined as any area defended against conspecifics (Burt 1943; Noble 1939), and on the other as "an exclusive area, not merely a 'defended' one" (Pitelka 1959). As noted by Bates (1970), the former interpretation stresses the behavioral corollaries of territoriality, while the latter focuses on the ecological significance of the existence of territories.

Taking Pitelka's definition of territoriality, all four study groups had a territory — and so have all primates studied to date. If territory is considered as a defended area, only a few primate species (e.g., *Hylobates lar* — Ellefeson 1968; *Callicebus moloch* — Mason 1968) — and two of the four study groups — can be described as *territorial*. Burt's definition of territoriality is used here, therefore, because it emphasizes an important variation in dispersal patterns within the Primate Order: "Although Pitelka's shift in emphasis from the behavioral implementation of spacing to its adaptive function is an important contribution, nevertheless the means of effecting this ecological exclusiveness remains an important problem in behavior" (Ripley 1967).

There is little justification for calling the two northern study groups *territorial* in Burt's sense of the word; with intergroup encounters taking place throughout extensive overlap areas between neighboring groups, there was no evidence that these encounters defined geographical boundaries of exclusively used areas. In contrast, the behavior of the southern study groups and groups at Berenty can reasonably be described as territorial; overlap between the ranges of neighboring groups was minimal and, although experimental substantiation is lacking, I believe that encounters in this narrow overlap area did serve to redefine and defend the boundaries of tangible areas of forest. Certainly, the animals' "interpretation" of intergroup encounters differed between areas; in the north, animals might continue on their initial route after an encounter, even if that route was taking them away from the center of their range and toward the exclusive area within the encountered group's range. In the south, animals always moved away from the edge of their range toward its center after an encounter, whether they were apparent victors or not.

If one turns to the movement of animals around their range, there was seasonal as well as regional variation. In both study areas, animals moved further in the wet season than in the dry. They entered more quadrats each day as they did so, although home-range size itself did not increase; animals simply covered their range faster in the wet season. All four study groups tended to concentrate their time in core areas, particularly during the dry season. These general and seasonal characteristics of ranging were common to both the northern and southern study areas, but regional differences were present nonetheless; in both seasons, the northern groups moved farther than the southern groups. In the north, core areas were commonly used by neighboring groups as well as by the resident group; in the south, as in most primate species (Bates 1970), core areas were used almost exclusively by the resident group. Repeated sampling of Group III in June 1974 suggests that, although this group did each day move farther in 1974 than in 1971, the ranging patterns described here may in general be highly stable over long periods of time.

In sum, previous assumptions are no longer tenable that *P. verreauxi* is territorial, with territory size varying between forests; one of the most striking results of the study was the discovery of a clear regional difference, ecological as well as behavioral, in the way animals partitioned their environment. This fundamental difference in pattern was associated with other, less-pronounced, regional variation in ranging patterns, and with seasonal trends that were similar in both study areas.

b) Feeding

P. verreauxi is vegetarian, and a high proportion of its diet consists of mature and immature leaves. Limited data suggest that feeding patterns remain stable over a period of years but, within these limits, there was marked variation through space (locally and regionally) and through time (seasonally). Considering first variation through space, local differences were primarily restricted to the species composition of the diet of the two neighboring groups under study in each area. Only eight of the twelve species most commonly eaten by each group in the north were the same, and a similar situation pertained in the south. This may have been due to differing traditions between groups, or to local differences in the availability and distribution of the various food species.

Regional differences were much more extensive; in comparison with the southern groups, animals in the north fed longer each day during the dry season, had a diet that was more varied and almost totally different in composition (although they fed on a smaller proportion of the total number of species present), and ate fewer fruit in the wet season and more immature leaves and fewer mature leaves in both seasons. Much of this variation was probably due to regional differences in the distribution and availability of foods, but the data nonetheless contain enigmas. For example, four tree species were found to be present in both forests; yet the frequency with which these trees were used as food, and the parts of the tree eaten, varied strikingly between study areas; frequency differences could not be explained as a function of differing abundances of the tree species in each forest. Possible explanations for this difference range from the simple to the highly complex. Differing food preferences among animals within and between study areas may account for some of the variation. However, in other instances, subtle shifts in the nutritional value of food may be critical; when a species was eaten commonly in one area and little or not at all in the other, another species providing equal or greater nutritional value may have provided a better alternative for these animals.

Turning to variation through time, seasonal changes in feeding behavior were clear in both study areas, and the direction of these changes was similar in both. In the south, however, changes were generally more pronounced, probably because of the greater climatic extremes to which the area was subjected. In comparison with the dry season, animals in the wet season fed for longer each day on a narrower and almost completely different range of food species. They ate more flowers, fruit and immature leaves and fewer mature leaves. Evidence (discussed below) indicates that in both study areas these seasonal changes were associated with, and reflected closely, seasonal cycles in the vegetation.

As with ranging behavior, patterns of feeding in *P. verreauxi* appear to be quite stable over several years. In both study areas there was overall a close positive correlation between daily distance moved and time spent feeding; thus, just as Group III moved farther each day in June 1974 than they had in June 1971, so they fed significantly longer. Otherwise, they ate many of the same foods for similar amounts of time. Furthermore, the total number of food species eaten was almost identical in the two years. In both, flowers and immature leaves formed a negligible portion of their diet, but in 1974 animals ate significant quantities of bark and almost no fruit, while in 1971 they had eaten almost no bark and a high proportion of fruit; this change was almost entirely due to the replacement of the fruit of a *Grewia* sp., which was their second most preferred food in 1971, by the bark of *Cedrelopsis grevei,* which was their second most-preferred food in 1974. Mature leaves also formed a somewhat higher percentage of animals' diets in 1974. Since no analysis of vegetation was undertaken in 1974, little can be said concerning the possible causes of this variation; a more evenly distributed pattern of rainfall in 1974 (Russell pers. comm.) was probably associated with some differences in the phenological cycle of tree species in the forest, which would, in turn, have affected the pattern of feeding of the animals.

c) Locomotion

The results of a limited analysis of locomotion were similar for both study areas. All animals tended to assume a vertical position when moving, and 75% of the time they spent feeding was in either a vertically clinging or sitting posture. Nonetheless, in both study areas the greatest diversity of postures occurred when animals were feeding. Animals used all substrate categories regularly except the ground; feeding behavior occurred predominantly among smaller branches and twigs. The similarity of results for both forests contrasts with the considerable differences apparent in their physical structure, but no systematic analysis of substrate availability in each forest was made; only in the light of such an analysis could the full significance of the results be understood. Data from the analysis of vegetation did indicate a much greater horizontal component in the northern forest; it can thus be tentatively concluded that animals were highly selective in their choice of substrate and posture and that the nature of this selectivity differed little between forests.

d) Daily activity patterns

The daily pattern of activities was indistinguishable between neighboring groups, and similar in both study areas. Seasonal changes in this pattern were striking and, again, similar in both study areas. In the dry season, activity only began an hour or two after sunrise, with sunning behavior. This might last for over an hour before the group moved off to feed. Animals fed more or less continuously until early afternoon, when they took up their sleeping positions for the night. In contrast, in the wet season animals were usually feeding before dawn, and most feeding activity had ceased by mid-morning. After a prolonged siesta animals began feeding again in the late afternoon, and continued to feed until well after sunset. In both study areas, more time was spent in the shade during the wet season than during the cooler dry season.

2. Social Behavior

Within-group interactions, whether agonistic or nonagonistic, were infrequent and did not appear to be highly complex; a feeding hierarchy could be discerned in all four groups studied, so that the outcome of an agonistic encounter over a food source between any two members of each group could usually be predicted with accuracy. Yet this simplicity of within-group social structure relative to many other primate species (e.g., *Pan troglodytes,* Van Lawick Goodall 1968; *Papio hamadryas,* Kummer 1967, 1968; *Lemur catta,* Budnitz and Dainis 1975, Jolly 1966) does not imply that the four groups were each a social replica of the other; group composition varied widely, and this had implications for the nature and frequency of interactions taking place within the group. For example, the presence of a mother and her newly born infant tended to increase the frequency of aggression as the mother repeatedly fended off the approaches of other members of the group. The presence of one or more immature animals increased the frequency of play behavior within a group, and immature animals also tended to have a specially intense grooming relationship with an adult female. Further, distinctive, "idiosyncratic" relationships existed between animals, such as that between the two adult males in Group III. The variation observed could not be classified as "local" or "regional"; groups varied, apparently without respect to locale.

Seasonal changes in social interaction occurred in both study areas. Some of these changes were contingent upon seasonally occurring events in the animals' annual reproductive cycle, namely, the mating season and the birth season. Others were associated with more general changes taking place in accordance with overall environmental change between seasons; for example, animals fed longer in the wet season, and the frequency of agonistic behavior increased at that time— a predictable result, since most aggression occurred within a feeding context. Again, play behavior was seen only in the wet season in both study areas, probably because of energy limitations in the dry season (see below).

3. Correlates of Behavioral Variability

This study compared the behavior and ecology of four groups belonging to the species *P. verreauxi.* The comparison was between groups living in different habitats and between the behavior of each of those groups in the wet season and the dry season. The nature, duration, and timing of many behaviors were found to vary both regionally and seasonally. This variation appeared to be the result of a complex interaction between behavioral and environmental factors; although substantiating data are not available, certain aspects of *P. verreauxi*'s behavior can probably best be understood as a response to physiological requirements that must be met in the face of differing environmental conditions. Data were collected on two aspects of the environment, namely climate and vegetation, which appeared to be important in this regard, and correlations were found between variations in these and the behavioral data. Although causality can be neither assumed from nor demonstrated by these correlations, they do provide a useful basis on which to formulate hypotheses concerning the relationship between behavior and ecology. These hypotheses will be tested in future studies.

a) *Temperature and day length*

Mean maxima and minima of temperature varied seasonally in both study areas; temperatures were higher in the wet season ($39.5°C$ and $18°C$ in the north, $39°C$ and $16°C$ in the south) than in the dry season ($31.5°C$ and $14°C$ in the north, $36°C$ and $8°C$ in the south). During the dry season, particularly in the south, animals were exposed to prolonged periods of low ambient temperature at night. Observational data include a series of behaviors that may have contributed to the maintenance of animals' body temperature despite the daily extremes of high and low temperature.

During the dry season, animals spent more time exposed to direct sunlight than in the wet season. Three extrinsic factors may have been partly responsible for this. First, loss of foliage by many trees during the dry season in both study areas must have reduced the overall availability of shade to the animals. Second, the total number of hours of sunshine was greater during the dry season because there were fewer overcast days during that period (although day length was itself shorter). Third, the location of food sources changed between seasons and may have resulted in animals spending more time in parts of trees exposed to sunlight; in the north, for example, animals in the dry season spent more time feeding in high emergents where shade was minimal. Differences in insolation between different times of day could similarly be explained as incidental consequences of differences in the location of feeding or locomotor activities at these times.

In addition to these explanations, the following hypotheses are presented: animals sought or avoided the sunshine according to the level of ambient temperature and their own body temperature; they exhibited sunning behavior, particularly in the early morning during the dry season, in order to counter the cooling effects of low night-time ambient temperatures; and they huddled together at night to minimize individual heat loss. In otherwords, it is argued that a set of behaviors with thermoregulatory functions has evolved in *P. verreauxi* in response to environmental fluctuations in temperature.

Rainey (1970) demonstrated experimentally that the rock hyrax *(Heterohyrax brucei)* must use behavioral as well as physiological mechanisms to maintain its body temperature. Behavioral mechanisms include huddling at night (thereby reducing the total heat loss of the two animals involved by 30%), increasing or decreasing body contact with rock substrates (depending on ambient and internal temperature), and sunning behavior (through which an animal could raise its temperature by 4°C in one hour). Comparable research has not been done on any primate species, although a limited investigation of the relation of rectal temperature to ambient temperature fluctuations in various captive prosimians has been made (Bourliere et al. 1953, 1956); animals under captive conditions, particularly members of the smaller species, did not maintain their body temperature when exposed to a range of ambient temperatures comparable to that in a natural environment. The biological significance of these results is unclear, however, and *P. verreauxi* was not among the species studied.

In comparison with the wet season, animals in the dry season rose later in the morning, moved less, fed less, and stopped playing and ceased activity for the day earlier. Low ambient temperatures at night in the dry season may provide a partial explanation for these seasonal changes in general activity as well as for the more specific behaviors described above. It is argued that the late start of the day's activities in the dry season was in part a result of the animals' need first to raise their body temperature; animals became active only when the ambient temperature had risen several degrees from its dawn level and they had sunned themselves for an hour or more. Even unhabituated animals were sluggish prior to this point and neither moved off nor responded strongly to my presence as they did at other times of day. Further, it is possible that higher levels of energy expenditure on thermoregulation during the dry season had more general and pervasive effects; if these higher levels had the net effect of reducing energy available for other activities, then the observed overall reduction in activity and, particularly, the cessation of play behavior would be indirectly attributable to the thermoregulatory stresses to which animals were exposed at that time.

It should be stressed that these hypotheses remain untested and that they do not suggest a simple, direct relationship between temperature and activity level in *P. verreauxi* in particular or primates in general: activity levels represent the compromise resulting from a spectrum of interacting extrinsic and intrinsic factors, and studies of *Papio ursinus* (Stoltz and Saaman 1970) and *Colobus badius* (Clutton-Brock 1972) have indeed shown an inverse relationship between temperature and daily range. It is argued, rather, that the wide daily temperature range in the dry season constitutes a thermoregulatory stress for *P. verreauxi* and is one of a series of constraints placed on its activity level at that time.

Day length in both study areas varied by about three hours between seasons. Tattersall and Sussman (1975) have found that light levels are probably a decisive factor in the control of activity patterns in *Lemur mongoz,* and Pariente (1974) has shown that light levels trigger the onset and cessation of activity in *Lepilemur mustelinus* and *Phaner furcifer.* It is unlikely that light levels have a direct activating effect on *P. verreauxi* or that the increase in day length in the wet season was responsible for the longer daily period of activity at that time; in the wet season, animals often moved and fed before sunrise and after sunset, while in the dry season they were generally inactive for two or three hours after sunrise and before sunset. Insofar as animals were almost never active between 1930 h and 0430 h, when light levels were at their lowest point, these levels may be said to influence their behavior; within these limits, animals appeared to become active or cease activity in response to cues other than light.

b) Vegetation

Having spent several years studying the ecology of baboons in East Africa, Altmann and Altmann (1970) concluded that the distribution of vital resources such as water had an important effect on ranging patterns and home-range size among these animals. They further postulated that:

> For any set of tolerable ecological conditions, the adaptive activities of baboons tend in the long run toward some optimal distribution away from which mortality rate is higher, or reproductive rate is lower, or both.

Their approach, formulating testable hypotheses concerning the relationship between baboon strategies and their environment, goes beyond the "simple correlation combined with *post hoc* explanation" of earlier studies (Altmann 1974); in addition to postulating causal relationships based on observed correlations, they reformulate these postulates as hypotheses that can then be tested separately. The concept of the *strategy* is fundamental to this approach: it is assumed that animals act as if they were obeying a set of rules governing their behavior and it is the goal of the researcher to determine these "rules" by constructing a series of models with varying sets of rules and testing the results generated by them against the "real life" behavior of the animals.

Underlying the assumptions that animals organize. their activities, and that they do so in a way that is detectable to the observer within a short period such as a year, are a further series concerning the "goals" sought   by the animals through these strategies. Thus we find *reproductive strategies* by which animals seek to maximize their reproductive success (Trivers 1972) and, particularly pertinent to this discussion, *feeding strategies*. In a review paper, Schoener (1971) defines the aim of a theory of feeding strategies as being ". . . to specify for a given animal that complex of behavior and morphology best suited to gather food energy in a particular environment." With few exceptions, (e.g., Botkin et al. 1974), the models built by ecologists (which have provided a great impetus to primate research) have all been based on the same premise: the goal of an animal or group's feeding strategy is to optimize its energy input (Emlen 1968; Levins and MacArthur 1969; MacArthur and Pianka 1966). Increasingly, however, this is being questioned and the importance of other factors such as nutrient balance (Boyd and Goodyear 1971) and plant toxicity (Freeland and Janzen 1974) is being stressed.

The term *strategy* is used in this discussion where other researchers have themselves used it, with the added caveat that it involves a number of generally unspecified assumptions. My general approach at the time of this study was one of "simple correlation. . . ." and I can present no evidence of a "strategy" for *P. verreauxi*. At best, a number of patterns of behavior emerged from the study that accord well with some of the predictions of the ecological models. However, neither the mathematical significance of this apparent "fit" nor the "goals" of the observed behaviors are known, and no test has been made of the extent, if any, to which these behaviors reflected a strategy, or underlying set of rules governing their patterning in the long term. In the following discussion, therefore, predictions and theories arising from ecological research are used only to provide a framework for preliminary interpretation of the variation found in feeding behavior in this study and a base from which to consider hypotheses concerning the relationship between this variation and variation through space and time in the animals' habitat.

Sussman (in press a) found that the feeding behavior of *Lemur fulvus* and *Lemur catta* accorded well with at least three of the predictions arising out of feeding-strategy models. These were, first, that animals utilizing abundant and predictable food resources should choose a small range of food items relative to the total range of items present; second, that when the absolute abundance of food is low, animals should show deceased selectivity; and third, that with low food productivity and patchy resources that are presumably less predictably distributed for the animals in question, home-range size should increase. *Lemur fulvus* live in continuous canopy forest, in small overlapping ranges (0.75-1.0 ha), and the major proportion of their diet consists of kily leaves and fruit, which are abundant in the forest. *Lemur catta* live in large home-ranges that overlap little (6 ha — Berenty, 8.8 ha — Antserananomby, up to 23 ha in more arid areas), and occupy a variety of forest types, including very dry brush and scrub forests. They eat a wider variety of foods, feeding on two to three times as many species of plants as *L. fulvus*. Further, while *L. fulvus* can apparently survive on a diet consisting almost entirely of mature leaves at certain times of the year, *L. catta* ranges widely horizontally and vertically to find flowers, fruits, and immature leaves. Sussman's hypothesis, thus, was that *L. fulvus* uses a strategy that involves moving in a small range, feeding on a few, abundant resources; *L. catta's* strategy, in contrast, is to move over a more extensive area, feeding on a wide variety of less commonly occurring and patchily distributed foods. Comparative data suggest that these strategies may be characteristic of the whole species and not just of the populations studied intensively by Sussman (Sussman pers. comm.).

Data on the distribution of resources and the feeding behavior of *P. verreauxi* suggest a pattern intermediate between those of *L. fulvus* and *L. catta*. Confirming Jolly's description (1966) of *P. verreauxi* as a "jack-of-all-trades," its feeding behavior also appears to be more flexible than that of the two lemur species. First, however, a summary is given of the evidence from which the hypothesis concerning their feeding pattern was formulated.

Considering the distribution and availability of plant species and hence of potential foods, there were similarities and differences between the two areas. In both, the profile of the forest was relatively low, and trees were generally narrow-girthed and present in high densities; in both, most tree species were rarely occurring and widely dispersed; phenological cycles were closely synchronized among species within each forest, so that new leaves, flowers, and fruit were present in abundance only seasonally. In the north, rainfall was higher and more evenly distributed around the year than in the south, where little rain fell and this only in three months of the year. Associated with these differences, vegetation in the north was more diverse (by species composition), and more abundant and lush than in the south; leaf-size was generally greater, lianas were more plentiful, the canopy was well developed and ground cover prolific.

If one turns next to the feeding and ranging behavior of the four groups studied in these habitats, seasonal and regional variations were found in these as well as in many other aspects of their behavior. The direction of seasonal changes was generally the same in both study areas, although they tended to be more pronounced in the south. In the dry season in both study areas, animals moved short distances each day and fed for short periods on a wide variety of food species, with leaves constituting a major proportion of their diet. In the wet season they moved farther each day, and fed for longer periods on fewer species, with flowers and fruits predominant in their diet. Home-range size was similar in both study areas, but the ranges of groups in the north overlapped extensively, whereas in the south animals lived in almost exclusive ranges, which they appeared to defend against intrusion by neighboring groups. All groups used a few part of their range more frequently and for longer periods than the rest of it.

Using these data, we can construct a hypothesis concerning the feeding patterns in each study area. In the north the scattered distribution of the many rare food species included in the animals' diet determined the minimum area over which the group must range in the course of a year. Animals covered this range throughout the year in order to monitor these scattered food sources and eat them as they became available; most of these foods were eaten seasonally only, and many were eaten in very small quantities. As a consequence, at any one time the *total* food available to animals within the range was in excess of the group's immediate needs and overlap between groups could therefore be extensive. Intergroup encounters functioned to preserve group integrity, and on occasion, to dispute access to a favored food source.

In contrast, in the south the dispersion of groups was determined more by *total* food availability at certain times of year rather than by the distribution of critical but scattered food sources; the whole range was necessary to support the group toward the end of the dry season, when total food availability was low. It is argued that round-the-year maintenance of a largely exclusive range was a more efficient means of assuring this seasonally crucial resource area than seasonal conflicts to establish exclusivity when total food availability was low.

In neither study area, thus, did animals primarily eat abundant and predictable food resources as *L. fulvus* did, but neither did they monitor so wide an area as *L. catta*. Rather, they fed on a wide range of plant species, some of which were abundant but many of which were widely and unevenly scattered through the forest. Both seasonal and regional variations in this overall pattern accord well with the predictions of the foraging strategy models; in both study areas, animals fed on a wider range of food species in the dry season when the abundance of food was lower. In the south, which appeared to have lower overall productivity and more patchy, less predictable resources than the north, animals ate a proportionately wider range of food species relative to the total number of tree species present (although absolutely less); although the size of each group's home-range was no greater than in the north, overlap between ranges was less and groups apparently defended their home-range; the ratio of biomass to area was thus effectively lowered. The presence of this defensive behavior in the south fulfilled another of the feeding strategy models' predictions: a lowering of food density will favor switching from lack of defense to defense, if the area in question is not too large (Schoener 1971). It is of interest to note that *L. catta*, living in a larger range with low overall abundance and patchy distribution of food, does not defend its range, although each group has exclusive use of most of its range. Groups do, however, advertise their presence with "loud-calls." It could be argued, then, that this further substantiates the prediction: living in ranges too large to defend economically, yet requiring largely exclusive use of these ranges, *L. catta* evolved a loud-call which functions as a low energy-consuming spacing mechanism. In Table 7.25, data on the feeding and ranging behavior of the three species are presented in summary form to illustrate further these relationships and predictions.

Much remains unexplained. One problem concerns the animals' tendency to feed only briefly on what appeared to be extensive food resources; animals often ranged over wide distances to feed briefly on some part of a large but rarely occurring tree. It has been argued elsewhere (Richard in press) that these foods may have contained critical trace nutrients and needed to be eaten only in small quantities. However, recent research has indicated that various secondary compounds synthesized by plant species may have toxic effects on herbivores ingesting them. As with all potential toxins, a dosage effect is found (Freeland and Janzen 1974), and the authors go on to discuss the implications of this:

TABLE 7.25

POSTULATED RELATIONSHIPS BETWEEN DISPERSION AND SIZE OF FOOD SOURCES, AND HOME-RANGE SIZE AND OVERLAP BETWEEN GROUPS, IN *P. VERREAUXI, LEMUR FULVUS, AND L. CATTA*

Area	Distribution and Size of Food Sources	Diversity of Diet	Food Availability within Home-Range	Home-Range Size	Degree of Overlap	Defense of Home-Range
P. verreauxi Ampijoroa (northern study area)	Some food sources widely scattered and small	high	In excess of group's requirements at any given time	large	extensive	No
Hazafotsy (southern study area)	Many food sources widely scattered and small	high	Becomes limiting factor during critical periods e.g., drought	large	minimal	Yes
Lemur fulvus Antserananomby & Tongobato	Major food sources concentrated and large	low	Seasonally in excess of group's requirements	very small	seasonally extensive	No
L. catta Berenty & Antserananomby	Major food sources very widely scattered and small	very high	Sufficent for resident group's requirements	very large	minimal	No

> Consuming plant secondary compounds is a potentially dangerous and metabolically expensive enterprise. . . . Generalist herbivores are forced to eat small amounts of several kinds of food. The total of these limited amounts that can be eaten at any one time is probably less than the total amount of these foods available.

Future research will investigate the possibility that the diversity of *P. verreauxi*'s diet and the frequent failure of animals to eat apparent food present in large quantities are due to the species' inability to detoxify many secondary compounds.

c) Other factors

Parameters of which no account was taken in this study may also have been important influences on the behavior patterns of the four groups. Schoener (1974), for example, has discussed the effects of competition on patterns of resource utilization among a number of sets of sympatric species. In the northern study area, *P. verreauxi* was sympatric with six other prosimian species in addition to other mammals and many bird and reptile species. In the south, where productivity and overall biomass appeared lower, animals seemed to have fewer potential competitors — but there was less to compete for. No parallel studies of other species were carried out, however, and the influence of other species as competitors or predators is unknown.

8
The Relationship between Ecology and Social Organization

This study aimed at documenting the ecology and social organization of a prosimian species, with particular emphasis upon its ability to demonstrate behavioral flexibility. A further goal was to provide comparative data that might shed light on the general relationship between ecology and social organization. In the concluding discussion below, *P. verreauxi* is placed in context as a leaf-eating primate and its ecology and social organization are compared and contrasted with those of other leaf-eaters. There follows a broader consideration of classificatory systems that have attempted to categorize different sets of relationships between social organization and ecology in a spectrum of primate societies. Finally, changes of emphasis and approach currently taking place in primate field studies are reviewed briefly, and some possible implications of these changes for our understanding of socioecology are considered.

1. *P. verreauxi* as a Leaf-Eating Primate

A folivorous diet is one that contains a high proportion of leaves. Such a definition is open to a wide range of interpretations, as Eisenberg (in press) has noted. Despite this vagueness, with few exceptions

183

there is general agreement in the literature as to which primates deserve the label *folivore*. In one study, a folivorous diet was operationally defined as containing by weight 45% or more leaves, buds, leaf shoots, bark, pith, and unripe fruit (Kay 1973). The list of primates whose diet met this definition differs little from other less systematic classifications. In other words, the leaf-eaters form an unusually distinctive group. Which species belong to this group? What are their distinguishing characteristics? And does *P. verreauxi* show such characteristics?

Leaf-eating primates are found in all three areas in which major primate radiations have occurred, namely South America, Africa and Asia, and Madagascar. In the New World, a leaf-eating niche is occupied by the howler monkey, *Alouatta* spp. (Hladik and Hladik 1969). In Africa and Asia, Colobines are the folivores *par excellence*; but a high proportion of the gorilla's diet consists of leaves and shoots and when more data are available other apes, such as the orang utan, *Pongo pygmaeus*, may deserve inclusion in this dietary category. (Chivers 1974; Clutton-Brock 1972; Hladik and Hladik 1972; MacKinnon 1974; Rodman in press; Schaller 1963; Struhsaker 1975 and in press). The Indriids and two lemur species, *Hapalemur griseus* and *Lepilemur mustelinus* are the extant leaf-eaters of Madagascar (Petter 1962a), but morphological evidence suggests that at least one group of recently extinct lemurs, the Megaladapinae (a lemurid subfamily containing three species), should also be included (Tattersall 1975).

Compared with a frugivorous one, a folivorous diet has a substantial fiber content: it is high in structural carbohydrates such as cellulose, hemicellulose, and related polymers such as lignin (Parra in press). These carbohydrates are less efficiently used, and often contain much more water than the nonstructural carbohydrates (sugars, starches, fructosans) that form the primary constituents of a frugivorous diet of ripe fruit, nuts, sap, and resin (Boyd and Goodyear 1971; Hladik et al. 1971). Thus folivorous primates must prepare their food thoroughly to maximize energy extraction, and must also ingest relatively large amounts in order to meet minimum energy requirements.

Given the exigencies of such a diet, the folivorous primates as a group show a number of morphological and behavioral traits that allow them to survive without taking in other, more readily assimilable foods. First, they tend to be large relative to primates with a more frugivorous diet; below a minimum body size, they could not ingest sufficient bulk to meet their energy requirements (Kay 1973).

Second, they have "more than average expression of molar tooth features associated with shearing, crushing and grinding" (Kay 1973; Kay and Hylander in press). Third, specializations of the digestive tract are present in most leaf-eaters; Hladik (1967) has shown that the relative surface area of the stomach and/or large intestine is generally much greater than in primates with a low-fiber-content diet. Where data are available, these morphological features appear to be as characteristic of *Propithecus*, and the Indriids in general, as they are of the other folivores (Hladik 1967; Kay and Hylander in press; Kay 1973; Napier and Napier 1967; Pollock 1975b).

Patterns of activity have been well documented for many mammals. Comparisons between studies are difficult to make because of differences in methods of data collection, but at a general level gross differences in patterns are found frequently to correlate with broad differences in diet. Diurnal insectivores and carnivores generally spend more time foraging and less resting than herbivores. Frugivores, in turn, tend to spend longer periods foraging each day than folivores. For example, the frugivorous/insectivorous coati, *Nasua narica*, spends most of the day foraging, with only occasional rest periods (Kaufmann 1962). Neal (1970) reports that packs of the banded mongoose, *Mungos mungo*, " . . . covered large distances, foraging as they went. . . . " In contrast, ungulates feed for shorter periods and show a daily bimodal distribution of feeding activity. The male Uganda cob spends 31-43% of its time feeding, depending on its social status (Leuthold 1966); the warthog, a browser, spends 20-40% of its time feeding (Clough and Hessam 1970).

A similar dichotomy is found within the Order Primates. The fruit- and insect-eating *Cebus capuchinus*, for example, spends a large proportion of daylight hours foraging (Hladik and Hladik 1969), compared with 20-25% for sympatric *Alouatta palliata* (Chivers 1969). In Africa, savanna-living, omnivorous baboons spend most of the day foraging (Altmann and Altmann 1970; Stoltz and Saayman 1970), while the folivorous Colobines have morning and afternoon feeding bouts taking up 44% (*C. badius*) and 19.9% (*C. guereza*) of their time (Struhsaker and Oates 1975). In a series of studies on *Colobus guereza, Cercopithecus aethiops*, and *Papio anubis*, Rose (in press a and b) found that *C. guereza* spent 30% of total time feeding, *C. aethiops* 46.7%, and *P. anubis* 47.1%. When progression associated with feeding was also included, these percentages rose to 30.4%, 48.2% and, strikingly, 64.2% respectively. As with morphological characteristics, *P. verreauxi* falls into the folivorous group according to this behavioral criterion; animals fed for 24-37% of the day, with morning and afternoon feeding peaks.

At least two factors underlie this differentiation of activity pattern according to gross dietary categories. First, in general a leaf-eater's food resources are relatively more densely, evenly, and predictably distributed than those of a frugivore or insectivore. A leaf-eater does not have to hunt for scattered prey species who have their own cycle of activities, nor travel extensive distances monitoring trees for the presence or absence of fruit. Thus, the ratio of time spent foraging to time spent feeding is probably lower for a folivore. Second, throughout the mammals folivorous diets are associated with depressed basal metabolic rates and an overall reduction of energy budget (McNab 1974 and in press). McNab has offered two explanations for this:

1) The proportion of nonassimilable energy in leaves may be high, so that energy intake per day by an animal feeding on such foods is limited by the maximal bulk that can be processed in a day.

2) Trees combat predation by the production of compounds that are generally toxic to mammals and insects (Freeland and Janzen 1974). The low basal rates of folivores may reduce a) the intake of these compounds, and b) the cost of their subsequent detoxification. The latter is particularly important, since it often involves the addition of glucose molecules to the toxic molecules, thus reducing net energy intake.

On the subject of general patterns of ranging and intergroup spacing, a recent review contrasted folivores and frugivores as follows:

> Given comparable habitats, the frugivores have larger home ranges and move more widely during their daily activities than do folivores of an equivalent size class. . . . many folivores but only some of the frugivores employ troop or individual vocalizations in maintaining spacing between adjacent troops. (Eisenberg et al. 1972)

The first part of this statement accords well with the predictions of feeding-strategy models, and most of the evidence available now from comparative studies tends to support it. Note that in the examples below, corrections for differences in group biomass have not been made; pairs of species of approximately equivalent size were chosen, but the primary aim of the comparison is to show differences in range size per group, not densities or total biomass. *Cebus* and *Ateles* groups both range over wider areas than *Alouatta*, and in Africa *Pan* and *Cercocebus* have larger home-ranges than *Gorilla* and *Colobus* (Fossey 1974; Schaller 1963; Struhsaker and Oates 1975; Suzuki 1969; Waser and Floody 1974). The relationship also holds for *Presbytis senex* and *Macaca sinica* in Southeast Asia (Hladik and Hladik 1972). Finally, in Madagascar the frugivorous *L. catta* lives in substantially larger ranges at Berenty than folivorous *P. verreauxi* (Jolly 1966).

Eisenberg et al.'s second assertion concerning the use of loud-calls by folivores as spacing mechanisms is only partially supported by available data. It is true that many folivores (e.g., *Alouatta*, *Colobus guereza*, *Indri*, *Presbytis* spp.) have loud-calls. It is also possible that, given the depressed energy budget of folivores, intergroup avoidance through loud-calling may represent a less energy-consuming form of group spacing than repeated intergroup interactions. At the same time, it is also true that many frugivores (e.g., *Lemur catta*, *Hylobates lar*, *Pan troglodytes*, *Cercocebus* spp.) have loud-calls. Furthermore, the postulated spacing function of loud-calls has never been clearly demonstrated, and they may in fact represent a more complex system with multiple functions. For example, Ellefson (1968) notes that an adult male gibbon " . . . sometimes moves a hundred yards or more to make contact with another group that is giving morning calls . . . . the 'great-call' of the adult female may serve simultaneously as social communication in both intragroup and intergroup activity." Goodall (1965) comments that chimpanzees may react in a variety of ways to loud-calls given by members of another group: " . . . they may ignore the sound; they may glance casually or stare fixedly toward the other group and call out themselves; or they may respond by actively leaping through the branches of a tree. . . ."

In sum, it is generally possible to differentiate between folivore and frugivore ranging patterns, and *P. verreauxi* falls within the former grouping. Contrary to a number of assertions, loud-calls for intergroup spacing do not characterize the folivores as a group; these calls are found in both folivorous and frugivorous species and further investigation is needed before conclusions can be drawn concerning their function(s), and the processes whereby they function. The absence of a loud-call in *P. verreauxi* does not, therefore, separate this species from other folivores.

Turning to more manifestly "social" aspects of the behavior of folivores, their unity as a group breaks down further. Eisenberg et al. (1972) characterized the larger arboreal primates as tending toward "either a uni-male structure or an age-graded-male system, with folivores especially tending toward a uni-male organization." However, the range of information now available provides only limited support for this generalization. They suggested that the New World leaf-eater, *Alouatta*, exhibits an age-graded-male system at high population densities and a uni-male system under conditions of low density, but it is argued here that an age-graded-male system cannot be differentiated from a multi-male system on currently available evidence (i.e., the studies of Baldwin and Baldwin 1972; Carpenter 1934; Chivers 1969; Collias and Southwick 1952; Glander, in press). The folivores of Africa, diurnal, group-living species, spent a large proportion of foraging time looking for insects. Clearly then, there are real exceptions to the proposed correlation between "solitary" living and "insectivory." Eisenberg et al. (1972) suggest that solitary habits may result from animals being nocturnal, " . . . since the coordination of groups would be difficult. . . ." But a recent study of *Lemur mongoz* has indicated that this species, while nocturnal, lives in "family" groups that maintain contact by auditory signals (Tattersall and Sussman 1975). Further, there is evidence that in some diurnal species that live in rain forests where visibility is restricted, group members coordinate their movements primarily by vocalizations (Gautier-Hion 1970). It could be argued that, in general, contact vocalizations by nocturnal primates would draw the attention of predators to their location, but the vociferousness of the "solitary" *Lepilemur*, for example (McGeorge pers. comm.) runs counter to this idea.

The change from generally solitary nocturnal habits to diurnal social habits has been attributed to "a change from a diet requiring individual hunting to food sources often locally distributed and at which social responses allow congregation for exploitation in common" and the small size of social units in many forest frugivores to "limiting conditions of food supply occasioned by the relatively stable conditions of tropical rain forest" (Crook and Gartlan 1966). In 1970 Crook expanded this idea, arguing that the stable food supply in many forests is likely to produce a population equilibrium just below the "ceiling" imposed by the availability of food, and that selection may consequently favor "adaptations permitting increased environmental carrying capacity rather than a seasonal maximum recruitment common in less stable regions." In effect, he argues, these adaptations would reduce intra- and interspecific competition and increase the means of protection from predators. It was postulated that the small, territorial groups of many frugivores represent an adaptation of this kind, and that such a form of social organization would reduce both inter- and intragroup competition.

While this model might be applicable to some species (e.g., *Hylobates lar*), its general validity is questioned on two grounds. First, there is mounting evidence that "gross, overall seasonal fluctuations" in food supply do occur in tropical rain forests (Richards 1966; Ellefson 1968; Medway 1970). Second, studies of the frugivorous *Pan troglodytes* (Suzuki 1969), *Ateles geoffroyi* (Eisenberg and Kuehn 1966), and *Cercopithecus mitis* (Aldrich-Blake 1970) suggest that their social organization is closer to the model proposed by Eisenberg et al. (1972) for frugivores. The latter argued that since frugivores needed to find ripening fruit trees within their home-range each day, " . . . the best feeding strategy involves breaking up the troop into small, independently foraging units that spread out to locate fruit trees within their home-range and then 'announce' the location of feeding spots."

Crook and Gartlan did not attribute a particular function to patterns of dispersion in folivores, but Eisenberg et al. stated that " . . . although primary folivores such as *Colobus guereza* and *Presbytis senex* eat considerable quantities of fruits, their feeding strategy is not predicated on a daily need of finding ripening fruit trees within their home-range. Small cohesive, uni-male social units are permitted within this strategy." However, recent studies of *Colobus badius* and *C. guereza* (Clutton-Brock 1974; Struhsaker and Oates 1975) have shown considerable dietary differences between these two closely related and reputed "leaf-eaters"; while *C. guereza* lives in small groups containing fewer than 14 individuals *C. badius* may live in groups containing up to 80 animals. It should also be noted that *Hylobates lar*, a "frugivore," lives in "small cohesive, uni-male social units."

Concerning the social organization of savanna-living primates, it has been argued that " . . . open country conditions of food supply and predation [favor] increase in group size . . ." (Crook and Gartlan 1966), yet Kummer (1971) noted that both hamadryas and gelada baboons often split up large herds into small one-male groups in order to forage during the day. The evidence concerning predation as a possible factor in selecting for increased group size is equally ambivalent. Both gelada and hamadryas baboons gather into large herds at night and sleep on cliffs, but patas monkeys live perennially in small one-male groups (Crook 1966; Hall 1965b; Kummer 1968). While the former apparently seek safety in numbers, the latter freeze and hide in the presence of danger, and the adult male exhibits constant watchful behavior and may give diversionary displays when the group is threatened. In the same paper Hall suggested that socionomic sex ratio rather than absolute group size may be critical for defending a group from predators. Goss-Custard et al. (1972) made a further point. If the proposed model holds good, " . . . one would expect some correlation between this type [multi-male troops] of social structure and the kind and severity of equally, present striking counterexamples. In Gombe National Park, *Colobus badius* was found in "large, multi-male troops of 40 or more animals," and in the Kibale Forest, Uganda, groups numbered from 19 to 80, and there was always more than one adult male in each group (Clutton-Brock 1974; Struhsaker and Oates 1975). Schaller's (1963) counts of group composition in the gorilla indicated an age-graded-male system, but Fossey's (1974) frequent sighting of groups containing up to three silver-backed males suggests a more variable pattern. Similarly, while *Colobus guereza* groups typically contain only one fully adult male, there are also reports of stable, multi-male groups (Oates pers. comm., Rose in press a). Turning to Asia, Jay (1965) found multi-male groups of *Presbytis entellus* in northern India, and Chivers (1973) reported *P. obscura* and *melalophos* having two to three adult males per group. Finally, in Madagascar *Indri indri* does live in the uni-male groups predicted by Eisenberg et al., but group composition is highly variable in *P. verreauxi*, groups containing from one to five adult males.

Clearly, there exists a wide and unexplained diversity of group size and composition among the folivores. Probably the most accurate statement concerning our ignorance of the determinants of social structure and the disconcerting lack of conformity in the social systems of forest primates grouped according to broad dietary critera has been made by Crook (1970):

> It seems likely that most leaf-eaters will turn out to live in one-male reproductive units. They share this characteristic nonetheless, with species of other tastes. Species living in multi-male troops are likewise diverse in feeding habits. . . . The size and composition of forest groups is doubtless programmed by a number of factors not all of which can be sensibly considered with our present knowledge.

In conclusion, insofar as the folivorous primates form a unified group, *P. verreauxi* belongs to that group. It has gut and tooth specializations, relatively large bulk, and a depressed-activity level and small range relative to animals with a higher proportion of fruit in their diet. All these characteristics can be related with greater or lesser specificity to the requirements and restrictions imposed by a diet high in structural carbohydrates and potentially toxic secondary compounds. Beyond showing that the depressed activity level of folivores extends to the frequency of interindividual interaction, current evidence does not point to a "type" social system for the folivores; while there may be a preponderance of one-male reproductive units, the studies cited above refute the idea of a close correlation between specific social systems and broad niche categories.

2. The Classification of Primate Social Systems

Although the distribution of nonhuman primates is almost completely limited to the tropics and subtropics, and although relatively few species have been studied in any detail, it is clear that there exists within the Order a tremendous diversity of social systems. Crook and Gartlan's paper in 1966 is the first and probably best known of numerous attempts to classify these diverse systems into a series of behavioral/ecological categories. The premise underlying this and similar classifications is that the social organization of a given species is adapted to, and determined by, the major features of its habitat and diet. The resulting schema have been criticized on various grounds, not least because " . . . viewed in the light of current knowledge, differences between groupings are less impressive than differences within them" (Clutton-Brock 1974).

In 1970 Crook modified the original classification, commenting: "Variations are probably best envisaged as continua in relation to gradients in environmental conditions, but for practical purposes we may use a preliminary categorization into five main types." The result differed from the 1966 classification in that it did not include the Prosimians, so that the *solitary* category was no longer necessary, and *population demes . . . with little social substructuring* was introduced as a new category. Each category was illustrated with fewer, carefully selected examples so that the overall classification was less open to such criticisms as Clutton-Brock's. Before an overall assessment of these classifications can be made, however, the validity of corollary arguments concerning the adaptive functions attributed to different types of social organization must be investigated.

Crook and Gartlan (1966) relate the "solitary" habits of the nocturnal primates to their insectivorous diet: insects provide a dispersed, mobile food source to which their predators respond with a dispersed form of social organization. However, at least one primate, *Lepilemur mustelinus*, is nocturnal, "solitary," and folivorous (Charles-Dominique and Hladik 1971). Thorington (1967) found that *Saimiri sciureus*, a diurnal, group-living species, spent a large proportion of foraging time looking for insects. Clearly then, there are real exceptions to the proposed correlation between "solitary" living and "insectivory." Eisenberg et al. (1972) suggest that solitary habits may result from animals being nocturnal, " . . . since the coordination of groups would be difficult. . . ." But a recent study of *Lemur mongoz* has indicated that this species, while nocturnal, lives in "family" groups that maintain contact by auditory signals (Tattersall and Sussman 1975). Further, there is evidence that in some diurnal species that live in rain forests where visibility is restricted, group members coordinate their movements primarily by vocalizations (Gautier-Hion 1970). It could be argued that, in general, contact vocalizations by nocturnal primates would draw the attention of predators to their location, but the vociferousness of the "solitary" *Lepilemur*, for example (McGeorge pers. comm.) runs counter to this idea.

The change from generally solitary nocturnal habits to diurnal social habits has been attributed to "a change from a diet requiring individual hunting to food sources often locally distributed and at which social responses allow congregation for exploitation in common" and the small size of social units in many forest frugivores to "limiting conditions of food supply occasioned by the relatively stable conditions of tropical rain forest" (Crook and Gartlan 1966). In 1970 Crook expanded this idea, arguing that the stable food supply in many forests is likely to produce a population equilibrium just below the "ceiling" imposed by the availability of food, and that selection may consequently favor "adaptations permitting increased environmental carrying capacity rather than a seasonal maximum recruitment common in less stable regions." In effect, he argues, these adaptations would reduce intra- and interspecific competition and increase the means of protection from predators. It was postulated that the small, territorial groups of many frugivores represent an adaptation of this kind, and that such a form of social organization would reduce both inter- and intragroup competition.

While this model might be applicable to some species (e.g., *Hylobates lar*), its general validity is questioned on two grounds. First, there is mounting evidence that "gross, overall seasonal fluctuations" in food supply do occur in tropical rain forests (Richards 1966; Ellefson 1968; Medway 1970). Second, studies of the frugivorous *Pan troglodytes*

(Suzuki 1969), *Ateles geoffroyi* (Eisenberg and Kuehn 1966), and *Cercopithecus mitis* (Aldrich-Blake 1970) suggest that their social organization is closer to the model proposed by Eisenberg et al. (1972) for frugivores. The latter argued that since frugivores needed to find ripening fruit trees within their home-range each day, " . . . the best feeding strategy involves breaking up the troop into small, independently foraging units that spread out to locate fruit trees within their home-range and then 'announce' the location of feeding spots."

Crook and Gartlan did not attribute a particular function to patterns of dispersion in folivores, but Eisenberg et al. stated that " . . . although primary folivores such as *Colobus guereza* and *Presbytis senex* eat considerable quantities of fruits, their feeding strategy is not predicated on a daily need of finding ripening fruit trees within their home-range. Small cohesive, uni-male social units are permitted within this strategy." However, recent studies of *Colobus badius* and *C. guereza* (Clutton-Brock 1974; Struhsaker and Oates 1975) have shown considerable dietary differences between these two closely related and reputed "leaf-eaters"; while *C. guereza* lives in small groups containing fewer than 14 individuals *C. badius* may live in groups containing up to 80 animals. It should also be noted that *Hylobates lar,* a "frugivore," lives in "small cohesive, uni-male social units."

Concerning the social organizaiton of savanna-living primates, it has been argued that " . . . open country conditions of food supply and predation [favor] increase in group size . . ." (Crook and Gartlan 1966), yet Krummer (1971) noted that both hamadryas and gelada baboons often split up large herds into small one-male groups in order to forage during the day. The evidence concerning predation as a possible factor in selecting for increased group size is equally ambivalent. Both gelada and hamadryas baboons gather into large herds at night and sleep on cliffs, but patas monkeys live perennially in small one-male groups (Crook 1966; Hall 1965b; Kummer 1968). While the former apparently seek safety in numbers, the latter freeze and hide in the presence of danger, and the adult male exhibits constant watchful behavior and may give diversionary displays when the group is threatened. In the same paper Hall suggested that socionomic sex ratio rather than absolute group size may be critical for defending a group from predators. Goss-Custard et al. (1972) made a further point. If the proposed model holds good, " . . . one would expect some correlation between this type [multi-male troops] of social structure and the kind and severity of predation. The presence of multi-male troops in open areas largely free of predators . . . does not accord with this expectation. . . ." Rowell

(1969), for example, found no evidence that predation contributed to overall mortality during a five-year study of three adjacent baboon troops in Uganda. (It must be noted that absence of predation cannot necessarily be taken as an absence of predation pressure. It could be a measure of the success of the social organization in minimizing effective predation, or of a recent reduction in predator density due to human interference. It could also mean that significant predation on a long-lived species cannot be measured in a period of five years.)

In conclusion, few would argue that these classification systems do not provide a useful overview of the range of variation in social organization present among the primates, and they have provided a powerful stimulus for research over the past ten years. However, many might question their current value in assessing the relationship between ecology and social organization; many of their predictions have not been fulfilled by the growing body of data, suggesting that there may be fundamental problems with the whole approach.

One such problem is the underlying assumption that social organization is to be determined simply by habitat and diet. A species' phylogenetic history may also be important in determining its response to a set of environmental conditions: " . . . different species tend to react to similar environmental pressures in different ways" (Clutton-Brock 1974). Struhsaker (1969) went further and emphasized phylogeny as a possible limiting factor: "In some cases, the immediate ecological variables may limit the expression of development of social structure and, with other species and circumstances, variables of phylogeny may be the limiting parameters."

Goss-Custard et al. (1972) have discussed another possible set of determining factors: social organization may reflect not only the *survival strategies* of individuals or groups but also their *reproductive strategies*. They argued that sexual selection could, under certain ecological conditions, result in a harem group structure, together with territoriality between groups or individuals. *Cercopithecus mitis* was cited as an example of a "territorial, one-male group" social system that both enhances adult survival and also permits a male to father several offspring at once and a female to increase the success with which she rears her young. An expanded version of this argument has been put forward in an extensive literature concerned with kin selection and the concept of *inclusive fitness* (e.g., Hamilton 1964; Trivers 1972; Wilson 1975).

A second problem concerns the assumption that the different elements of social organization consistently combine into a few, distinct, integrated systems. For, although rarely defined explicitly, the term *social organization* generally includes variables such as group size and composition by age and sex, stability of group membership, the identity of the primary reproductive unit (whether synonymous with, larger, or smaller than the group itself), degree of sexual dimorphism, patterns and mechanisms of disperison among neighboring groups. These variables are not, in fact, well correlated across species (Clutton-Brock 1974), so the use of only one as a classificatory criterion will group together species with social organizations disparate in other respects. A brief look at some details of social structure in *Erythrocebus patas, Papio hamadryas*, and *Theropithecus gelada*, which were grouped together by Crook and Gartlan in 1966, again by Crook in 1970, and by Eisenberg et al. in 1972, will illustrate this point.

All three species are found in grassland or arid savanna, and have similar vegetarian/omnivorous diets. Their populations subdivide into reproductive one-male units and there is marked dimorphism between the sexes. These similarities are matched by as many contrasts. The patas reproductive unit is also the social group, and although groups tend to space themselves through avoidance rather than direct confrontation, an adult male will chase away another group if it is close by. Extragroup males move alone or in small groups and have no contact with the one-male reproductive units. In the latter, aggression is infrequent and rarely results in actual contact between animals. Further, adult females more commonly direct aggression at the adult male than vice versa (Hall 1965b). In contrast, the reproductive units of geladas and hamadryas periodically coalesce into large herds, with adult males showing a high level of mutual tolerance.

However, even between these two species the resemblance is superficial:

> The one-male groups of gelada and hamadryas have in common the control and restriction of females by more dominant group members. This role, however, is monopolized by the male in hamadryas baboons; in geladas it is shared by the females. . . . The integrity of gelada one-male groups rests on the joint action of many. (Kummer 1971)

Kummer goes on to suggest that this difference may, in turn, be responsible for another; a gelada one-male group can spread throughout a troop, but a hamadryas group always retains its spatial unity and straying females are promptly retrieved and disciplined by the adult male. Further, subadult hamadryas males tend to adopt immature females or attach themselves to an adult male with a fully fledged harem, whereas subadult gelada males form all-male groups on the periphery of the herd, socially isolated from the harems. One final difference concerns the existence of "bands" in hamadryas social organization; geladas have no apparent social links between harem males, and the huge herds are amorphous gatherings of many one-male groups; members interact and sometimes challenge other bands, although more often two or three bands will coalesce to form a herd (Crook 1966; Kummer 1968; Kummer and Kury 1963).

In brief, such a classification tends to mask complexity rather than to explain it, and a similar problem arises in connection with the ecological categories used. Habitats are classified by blanket terms such as *forest, woodland savanna,* or *arid savanna,* and diets by *folivorous, insectivorous,* or *frugivorous.* Yet, as Clutton-Brock (1974) has pointed out, "Food supplies may be static or mobile, sparse or dense, heavily clumped or evenly dispersed, reasonably stable throughout the year or extremely variable." Similarly, studies of sympatric species are beginning to show that terms such as *folivore* may be of broad classificatory use but have little explanatory power and say almost nothing about niche; sympatric folivores have been found to feed on different food species and different food parts, and to have different ranging patterns and different group size, composition, and social structure (Clutton-Brock 1974; Hunt Curtin, and Chivers in press; Struhsaker and Oates 1975). Rudran (in press) has also noted that while larger frugivores like the mangabey, *Cercocebus albigena,* range over extensive areas throughout the year at a relatively high energy cost in order to find sufficient fruit, smaller monkeys such as *Cercopithecus mitis* live in smaller ranges and adopt a folivorous diet in time of fruit shortage: "The folivorous tendencies of blue monkeys are probably important in bringing about a substantial separation of their feeding niche from those of three other sympatric cercopithecines." Again, a retreat from simplistic terminology decreases, rather than increases, confusion.

3. The Study of Social Organization and Ecology

It is now generally recognized that statements such as "Species A is folivorous and forest-living" are of little predictive value for that species' social organization, and both the pragmatics and theory of socioecology are gradually changing as a result.

If ecological correlates of social organization are to be found, they must be sought at a more comprehensive and detailed level than has been used in the past. Analysis of environmental parameters must include the density, distribution, and availability through space and time of resources used by the animals. Records must be kept of climatic parameters such as temperature fluctuations, insolation, humidity, and rainfall. Similarly, social organization itself must be broken down and studied as an extended set of variables.

However, even if covariation is found between social and environmental variables, further research is needed if any understanding is to be reached of the interactive mechanisms inherent in such correlations. First, the nutritional value and potential toxicity of foods must be assessed in relation to the requirements and digestive efficiency of the species under study. The impact of climatic factors must be assessed in relation to the thermoregulatory abilities of the animals and an estimate made of the importance of these factors in the animals' total energy budget. Through comparative studies, the effects of differing levels of predation and competition from sympatric species can be investigated.

One of the primary implications of this need for research encompassing a range of disciplines, from field ecology to biochemistry, is that the days of the Lone Primatologist are largely numbered. While field studies conducted singlehanded have amassed an invaluable body of baseline information on the ecology and social organization of a spectrum of species, further insights into the complex interaction between animals and their environments can come only from the coordinated efforts of groups of researchers.

In a recent book review, Fleagle (in press) bemoaned the dearth of testable hypotheses generated by primate field research. In part, this has been due to the practical problems involved in manipulating what are often large, complex, highly arboreal, fleet-of-foot, and potentially dangerous animals in an experimental situation in an even larger forest. In part, hypotheses have been neither generated nor tested because field researchers were for many years content with the "simple correlation combined with *post hoc* explanations" approach (Altmann 1974). However, the new attention to detailed analysis of many aspects of a study species' behavior and ecology has both influenced and been influenced by a shift in underlying approach. Predictive models (e.g., Denham 1971) are now being formulated and, in a few instances, actually tested. Post (in prep.), for example, has constructed a model

simulating baboons' movements about their range on the basis of a
given set of resources in the environment, a given set of requirements
by the animals, and an "optimization strategy," or set of rules, that
the animals follow. (Obviously, the animals are acting "as if" following
this set of rules — the model does not purport to simulate their moti-
vation.) "Real" data were then collected on free-ranging baboons in
East Africa in order to test the model's ability to predict ranging
patterns in a real situation.

The predictive value of such models is likely to be only as great as
the underlying understanding of the factors and processes involved in
them. A truism, perhaps, but at a time when hypothesis-building and
testing are increasingly fashionable, this point deserves stressing. Indeed,
a long-term study of *Colobus badius*, which is probably the most
detailed of any study carried out to date, demonstrated no correlation
between ranging behavior and resource distribution (Struhsaker 1974).
From this it can be concluded, as Struhsaker does conclude, that
either the level of analysis was still insufficiently detailed to detect the
factors determining their ranging patterns or " . . . as a working hypothesis . . . the relatively diverse diet of red colobus monkeys permits
greater independence of their monthly ranging patterns from food
dispersion than for monkey species with a more monotonous diet."
It could equally be argued that seventeen months is an insufficient
time scale within which to detect the effects of a temporal constraint
such as natural selection.

We are left, thus, with a commitment to investigate many aspects of
the interactions of primates with their environments that are still
poorly understood, but with no certainty that the results will provide
us with a model whereby we can accurately predict social organization
from ecology. In reviewing the leaf-eaters as a group, it was clear
that the farther removed from the actual act of ingesting leaves the
comparison of their socioecology became, the greater the interspecific
variation became. This mounting variation may have been due not to
fine differences in the distribution of environmental resources and in
concomitant selective pressures, but to the freeing of increasingly
"social" aspects of behavior from selective pressure exerted by ecological factors. The final conclusions of this study are:

1) The evolutionary time scale is too great for many of the effects of
natural selection upon social organization to be understood and documented in even a "long-term" field study, and many of the apparent

relationships found in such studies are in fact no more than "noise" within the system.

2) At least some of the variability found in social organization represents not an immediate or even a long-term response to differing selective pressures exerted only by the environment, but a series of ecological equivalents arising out of a complex interplay of pressures such as kin and sexual selection acting throughout the phylogenetic history of every species.

3) Field research should focus not upon the general manner in which ecological pressures may have shaped primate societies, but upon the specific processes and mechanisms whereby animals exploit and interact with their environment and each other.

Bibliography

Aldrich-Blake, F. P. G. 1970. Problems of social structure in forest monkeys. In *Social behaviour in birds and mammals*, ed. J. H. Crook. Academic Press, New York. Pp. 79-102.

Altmann, J. 1974. Observational study of behavior: sampling methods. *Behaviour* 49: 227-67.

Altmann, S. A. 1962. A field study of the sociobiology of Rhesus Monkeys, *Macaca mulatta. Ann. N.Y. Acad. Sci.* 102(2): 338-435.

———. 1974. Baboons, space, time, and energy. *Amer. Zool.* 14: 221-48.

———, and Altmann, J. 1970. *Baboon ecology*. Chicago University Press.

———; and Altmann, J.; Hausfater, G.; and McCuskey, in press. The Life Cycle of Yellow Baboons. *Folia Primatol.*

Andrew, R. J. 1963a. Trends apparent in the evolution of vocalization in the Old World monkeys and apes. *Symp. Zool. Soc. Lond.* 10: 89-101.

———. 1963b. The origins and evolution of the calls and facial expressions of the primates. *Behaviour* 20: 1-109.

Baldwin, J. D., and Baldwin, J. I. 1972. Population density and use of space in howling monkeys *(Alouatta villosa)* in Southwestern Panama. *Primates* 13:371-79.

———. in press. Intertroop vocalizations of howling monkeys *(Alouatta palliata)* under conditions of high population density.

Basilewsky, G. 1965. Keeping and breeding Madagascan lemurs in capitivity. In *The international zoo yearbook*, vol. 5, ed. C. Jarvis, Zool. Soc. London.

Bates, B. C. 1970. Territorial behavior in primates: a review of recent field studies. *Primates* 11:271-84.

Bernstein, I. S. 1968. The lutong of Kuala Selangor. *Behaviour* 32: 1-16.

Bishop, A. 1962. Control of the hand in lower primates. *Ann. N. Y. Acad. Sci.* 102: 316-37.

Botkin, D. B.; Jordan, P. A.; Dominski, A. S.; Lowendorf, H. S.; and Hutchinson,

201

G. E. 1973. Sodium Dynamics in a Northern Ecosystem. *Proc. Nat. Acad. Sci. U.S.A.* 70(10): 2745-48.

Bourlière, F., and Petter-Rosseaux, A. 1953. L'homéothermie imparfaite de certains prosimiens. *C. R. des seances de la Soc. d. Biol.* 147: 1594.

———; Petter, J. -J.; and Petter-Rosseaux, A. 1956. Variabilité de la température centrale chez les lémuriens. *Mem. d. l'Inst. Sci. d. Madagascar* 10:303-4.

Boyd, C. E., and Goodyear, C. P. 1971. Nutritive quality of food in ecological systems. *Arch. Hydrobiol.* 69: 256-60.

Budnitz, N., and Dainis, K. 1975. *Lemur catta:* ecology and behavior. In *Lemur Biology*, ed. I. Tattersall and R. W. Sussman. Plenum Press, New York and London.

Buettner-Janusch, J., and Andrew, R. J. 1962. Use of the incisors by primates in grooming. *Amer. J. Phys. Anth.* 20: 129-32.

Burt, W. H. 1943. Territoriality and home range concepts as applied to mammals. *J. Mammal.* 27: 346-52.

Cabanis, Y.; Chabouis, L.; and Chabouis, F. 1970. *Végétaux et groupements végétaux de Madagascar et des Mascareignes,* vols. 1 and 2, Bureau pour le développement de la production agricole, Madagascar.

Carpenter, C. R. 1934. A field study of the social behaviour and social relations of howling monkeys *(Allouatta palliata). Comp. Psychol. Monogr.* 10: 168.

Carpenter, C. R. 1953. Grouping behavior of Howling Monkeys. *Extrait des Archives Neerlandaises de Zoologie.* 10(2): 45-50.

Cartmill, M. 1972. Arboreal adaptations and the origin of the order Primates. In *The functional and evolutionary biology of primates,* ed. R. Tuttle. Aldine Atherton Inc., New York.

Chalmers, N. R. 1968. Group composition, ecology, and daily activities of free living mangabeys in Uganda. *Folia Primatol.* 8: 247-62.

Charles-Dominique, P., and Hladik, C. M. 1971. Le *Lépilemur* du sud de Madagascar: écologie, alimentation et vie sociale. *La Terre et la Vie* 25(1): 3-66.

Chivers, D. J. 1969. On the daily behaviour and spacing of howling monkey groups. *Folia Primatol.* 10: 48-102.

———. 1973. An introduction to the socioecology of Malayan forest primates. In *Comparative Ecology and Behaviour of Primates,* ed. R. P. Michael and J. H. Crook. Academic Press, London and New York.

———. 1974. The Siamang in Malaya: a field study of a primate in tropical rain forest. *Contrib. Primatol.* 4, Karger, Basel.

Clough, G., and Hassam, A. G. 1970. A quantitative study of the daily activity of the warthog in the Queen Elizabeth National Park, Uganda. *E. Afr. Wildl. J.* 8:19-24.

Clutton-Brock, T. H. 1972. Feeding and ranging behaviour of the red colobus monkey. Ph.D. Disertation Univ. of Cambridge.

———. 1974. Primate social organization and ecology. *Nature* 250:539-42.

Crook, J. H. 1966. Gelada baboon herd structure and movement: A comparative report. *Symp. Zool. Soc. Lond.* 18:237-58.

——. 1970. The socio-ecology of primates. In *Social behaviour in birds and mammals,* ed. J. H. Crook, Academic Press, New York.

——, and Gartlan, J. S. 1966. Evolution of primate societies. *Nature* 210:1200-1203.

Denham, W. W. 1971. Energy relations and some basic properties of primate social organization, *Amer. Anthropol.* 73:77-95.

DeVore, I. 1963. A comparison of the ecology and behavior of monkeys and apes. In *Classification and human evolution,* ed. S. L. Washburn, Aldine, Chicago.

Ehrlich, P. R., and Raven, P. H. 1969. Differentiation of Populations, *Science* 165:1228-32.

Eisenberg, J. F. in press. The evolution of arboreal herbivores in the class Mammalia. In *The Ecology of Arboreal Folivores, Proc. Nat. Zool. Park Symp., No. 1,* ed. G. G. Montgomery. Smithsonian Institution Press, Washington, D.C.

Eisenberg, J. F.; Muckenhirn, N. A.; and Rudran, R. 1972. The relation between ecology and social structure in primates. *Science* 176:863-74.

Ellefson, J. O. 1968. Territorial behaviour in the common white-handed gibbon, *Hylobates lar* Linn. In *Primates: Studies in adaptation and variability,* ed. P. C. Jay. Holt, Rinehart and Winston, New York.

Emlen, J. M. 1968. Optimal choice in animals. *Amer. Nat.* 102:385-90.

Evans, C. S., and Goy, R. W. 1968. Social behaviour and reproductive cycles in captive ringtailed lemurs *(Lemur catta* L.). *J. Zool. Lond.* 156:181-97.

Flacourt, E. de. 1661. *Histoire de la Grande Isle de Madagascar.* Pierre L 'Amy, Paris.

Fleagle, J. in press. Review of *"The Simang in Malaya." Evolution,* (1976).

Fossey, D. 1974. Observations on the home range of one group of mountain gorillas *(Gorilla gorilla beringei). Anim. Behav.* 22:568-81.

Freeland, W. J., and Janzen, D. H. 1974. Strategies in herbivory by mammals: the role of plant secondary compounds. *Amer. Nat.* 108:269-89.

Gartlan, J. S., and Brain, C. K. 1968. Ecology and social variability in *Cercopithecus aethiops* and *C. mitis.* In *Primates: Studies in adaptation and variability,* ed. P. C. Jay. Holt, Rinehart and Winston, New York.

——, and Struhsaker, T. T. 1972. Polyspecific associations and niche separation of rain-forest anthropoids in Cameroon, West Africa. *J. Zool., Lond.* 168:221-66.

Gautier-Hion, A. 1970. L'organization sociale d'une bande de talapoins *(Miopithecus talapoin)* dans le Nord-Est du Gabon. *Folia Primatol.* 12:116-41.

Glander, K. in press. The feeding of howling monkeys and plant secondary compounds: a study of strategies. In *The Ecology of Arboreal Folivores, Proc. Nat. Zool. Park, No. 1,* ed. G. G. Montgomery. Smithsonian Institution Press, Washington, D.C.

Goodall, J. (van Lawick-). 1965. Chimpanzees of the Gombe Stream Reserve. In *Primate behaviour: Field studies of monkeys and apes,* ed. I. DeVore. Holt, Rinehart and Winston, New York.

Goss-Custard, J. D.; Dunbar, R. I. M; and Aldrich-Blake, F. P. G. 1972. Survival, mating and rearing strategies in the evolution of primate social structure. *Folia Primatol.* 17:1-19.

Hall, K. R. L. 1965a. Social organization of the Old World monkeys and apes. *Symp. Zool. Soc. Lond.* 14:265-89.

——. 1965b. Behaviour and ecology of the wild patas monkeys, *Erythrocebus patas,* in Uganda. *J. Zool. Soc. Lond.* 148:15-87.

——, and DeVore, I. 1965. Baboon social behaviour. In *Primate behaviour: Field studies of monkeys and apes,* ed. I. DeVore. Holt, Rinehart and Winston, New York.

Hamilton, W. D. 1964. The genetical evolution of social behavior. *J. Theoret. Biol.* 7:1-52.

Hausfater, G. 1972. Intergroup behavior of free-ranging rhesus monkeys *(Macaca mulatta). Folia primatol.* 18:78-107.

——. 1975. Dominance and reproduction in baboons *(Papio cynocephalus).* A Quantitative Analysis. *Contrib. Primatol.* 7, Karger, Basel.

Heimbuch, R., and Richard, A. F. An analysis of patterns of dispersion in a group of *Propithecus verreauxi.* In preparation.

Hill, W. C. O. 1953. *Primates: Comparative anatomy and taxonomy. l. Strepsirhini.* Edinburgh University Press.

Hinde, R. A. 1974. *Biological Bases of Human Social Behaviour.* McGraw-Hill, Inc., New York.

Hladik, A., and Hladik, C. M. 1969. Rapports trophiques entre végétation et primates dans la forêt de Barro Colorado (Panama). *La Terre et la Vie* 1:25-117.

Hladik, C. M. 1967. Surface relative du tractus digestif de quelques primates. Morphologie des villosités intestinales et correlations avec le régime alimentaire. *Mammalia* 31:120-46.

——, and Hladik, A. 1972. Disponibilités alimentaires et domaines vitaux des Primates à Ceylon. *La Terre et la Vie* 26(2):149-215.

——; Hladik, A.; Bousset, J.; Valdebouze, P.; Viroben, G.; and Delort-Laval, J. 1971. Le régime alimentaire des primates de l'Île de Barro Colorado (Panama) — Résultats des analyses quantitatives. *Folia Primatol.* 16:85-122.

Hunt Curtin, S., and Chivers, D. J. in press. Leaf-eating primates of Peninsula Malaya. In *The Ecology of Arboreal Folivores, Proc. Nat. Zool. Park Symp., No. 1,* ed. G. G. Montgomery. Smithsonian Institution Press, Washington, D.C.

Jay, P. 1965. The common langur of North India. In *Primate behaviour: Field studies of monkeys and apes,* ed. I. DeVore. Holt, Rinehart and Winston, New York.

Jewell, P. A. 1966. The concept of home range in mammals. *Symp. Zool. Soc. Lond.* 18:85-110.

Jolly, A. 1964a. Prosimians' manipulation of simple object problems. *Anim. Behav.* 12:560-70.

——. 1964b. Choice of cue in prosimian learning. *Anim. Behav.* 12:571-77.

——. 1966. *Lemur behavior.* Chicago University Press.

——. 1972. Troop continuity and troop spacing in *Propithecus verreauxi* and *Lemur catta* at Berenty (Madagascar). *Folia Primatol.* 17:321-34.

Kaufmann, J. H. 1962. Ecology and social behavior of the coati, *Nasua narica*, on Barro Colorado Island, Panama. *Univ. Calif. Publ. Zool.* 60:95-222.

Kay, R. F. 1973. Mastication, molar tooth structure and diet in primates, Ph.D. Dissertation, Yale University.

——, and Hylander, W. L. in press. The dental structure of mammalian folivores with special reference to primates and phalangeroids (Marsupialia). In *The Ecology of Arboreal Folivores, Proc. Nat. Zool. Park Symp., No. 1,* ed. G. G. Montgomery. Smithsonian Institution Press, Washington, D.C.

Kearney, T. H., and Schantz, H. L. 1911. The water economy of dry land crops. *Yearbook of Agriculture, No. 10.* Washington, U. S. Dept. of Agriculture.

Klopfer, P. H. 1972. Patterns of maternal care in three species of *Lemur:* II. Effects of early separation. *Z. Tierpsychol.* 30:277-96.

——, and Klopfer, M. S. 1970. Patterns of maternal care in three species of *Lemur:* I. Normative description. *Z. Tierpsychol.* 27:984-96.

Kummer, H., 1967. Tripartite relations in hamadryas baboons. In *Social communication among primates,* ed. S. A. Altmann. University of Chicago Press.

——. 1968. Social organization of hamadryas baboons, A field study. *Biblioteca Primatol.* 6, Karger, Basel/New York.

——. 1971. *Primate societies, group techniques of ecological adaptation.* Aldine-Atherton, Chicago and New York.

——, and Kurt, F., 1963. Social units of a free-living population of hamadryas baboons. *Folia Primatol.* 1:1-19.

Lawick-Goodall, J. van, 1968. The behavior of free-living chimpanzees in the Gombe Stream Reserve. *Anim. Behav. Monogr.* 1:165-311.

Leuthold, W., 1966. Variations in territorial behaviour of Uganda kob *Adenota thomasi* (Neumann 1896). *Behaviour* 28:214-57.

Levins, R., and MacArthur, R. 1969. An hypothesis to explain the incidence of monophagy. *Ecology* 50:910-11.

Lindburg, D. G. 1969. Rhesus monkeys: mating season mobility of adult males. *Science* 166:1176-78.

MacArthur, R. H., and Pianka, E. R. 1966. On optimal use of a patchy environment. *Amer. Nat.* 100:603-9.

McGinnis, P. 1973. Sexual behaviour of chimpanzees, Ph.D. Disertation, Cambridge University.

MacKinnon, J. 1974. The behaviour and ecology of wild orang-utans *(Pongo pygmaeus). Anim. Behav.* 22:3-74.

McNab, B. K. 1974. The energetics of endotherms. *Ohio J. Sci.*

——. in press. The energetics of arboreal folivores: Physiological problems and ecological consequences of feeding on an ubiquitous food source. In *The*

Ecology of Arboreal Folivores, Proc. Nat. Zool. Park Symp., No. 1, ed. G. G. Montgomery, Smithsonian Institution Press, Washington, D.C.

Marler, P. 1969. *Colobus guereza:* Territoriality and group composition. *Science* 163:93-95.

Martin, R. D. 1972a. Adaptive radiation and behaviour of the Malagasy Lemurs. *Phil. Trans. Roy. Soc. Lond. B* 264(862): 295-352.

Martin, R. D. 1972b. A Preliminary field study of the Lesser Mouse Lemur. *Z. Tierpsychol.* 9:43-89.

——; Doyle, G. A.; and Walker, A. C., eds. 1974. *Prosimian Biology.* Duckworth Ltd., London.

Mason, W. A. 1968. Use of space by *Callicebus* groups. In *Primates: Studies in adaptation and variability,* ed. P. C. Jay, Holt, Rinehart and Winston, New York.

Medway, Lord, 1970. The monkeys of Sundaland: Ecology and systematics of the Cercopithecids of a humid equatorial environment. In *Old World monkeys: Evolution, systematics and behavior,* ed. J. R. Napier and P. H. Napier, Academic Press, New York.

Milne-Edwards, A., and Grandidier, A. 1875, 1890-1896. *Histoire naturelle des mammifères: Histoire physique, naturelle, et politique de Madagascar.* vols. 6, 9, 10. Paris.

Napier, J. R. 1963. Brachiation and brachiators. In *The Primates,* ed. J. Napier and N. A. Barnicot, *Symp. Zool. Soc. Lond.* 10:183-95.

——, and Napier, P. H. 1967. *A Handbook of Living Primates.* Academic Press, London and New York.

——, and Walker, A. C., 1967. Vertical clinging and leaping: A newly recognized category of locomotor behavior of primates. *Folia Primatol.* 6:204-19.

Neal, E., 1970. The banded mongoose *Mungos mungo* Gmelin. *E. Afr. Wildl. J.* 8:63-71.

Nishida, T. 1968. The social group of wild chimpanzees in the Mahali mountains. *Primates* 9:167-227.

Noble, G. K. 1939. The role of dominance in the social life of birds. *Auk* 56:263-73.

Oates, J. 1974. The ecology and behaviour of the black-and-white colobus monkey *(Colobus guereza* Rüppell) in East Africa. Ph.D. Dissertaion, London University.

Oppenheimer, H. R. 1960. Adaptation of drought: Xerophytism. In *Plant-water relationships in arid and semi-arid conditions.* UNESCO Publ.

Pariente, G. 1974. Influence of light on the activitiy rhythms of two Malagasy lemurs: *Phaner furcifer* and *Lepilemur mustelinus leucopus.* In *Prosimian Biology,* ed. R. D. Martin, G. Doyle, A. C. Walker, Duckworth, London.

Parra, R. in press. Comparison of foregut and hindgut fermentation in herbivores. In *The Ecology of Arboreal Folivores, Proc. Nat. Zool. Park Symp., No. 1,* ed. G. G. Montgomery, Smithsonian Institution Press, Washington, D.C.

Petter, J. -J. 1962a. Recherches sur l'écologie et l'éthologie des Lémuriens malgaches. *Mem. du Mus. Nat. de l 'Hist. Naturelle, Ser. A.* 27:1-146.

——. 1962b. Ecological and behavioural studies of Madagascar lemurs in the field. *Ann. N.Y. Acad. Sci.* 102:267-81.

——. 1962c. Ecologie et éthologie comparées des Lémuriens Malgaches. *La Terre et la Vie* 109:394-416.

——. 1965. The lemurs of Madagascar. In *Primate behavior: Field studies of monkeys and apes,* ed. I. DeVore. Holt, Rinehart and Winston, New York.

——, and Peyrieras, A., 1974. A study of the population density and home range of *Indri indri* in Madagascar. In *Prosimian Biology,* ed. R. D. Martin, G. A. Doyle, and A. C. Walker. Duckworth, London.

——. Schilling, A., and Pariente, G. 1971. Observations éco-éthologiques sur deux Lémuriens malgaches nocturnes: *Phaner furcifer* et *Microcebus coquereli. Terre et Vie* 25:287-327.

Petter-Rousseaux, A., 1962. Recherches sur la biologie de la réproduction des primates inférieurs. *Mammalia* 26(Suppl. 1): 1-88.

——. 1964. Reproductive physiology and behaviour of the Lemuroidea. In *Evolutionary and genetic biology of primates,* ed. J. Buettner-Janusch. vol. 2. Academic Press, New York.

——. 1969. Day length influence of breeding season in mouse lemurs. Presented at Eleventh Ethological Congress, 1969.

Pitelka, F. A. 1959. Numbers, breeding schedule, and territoriality in pectoral sandpipers on Northern Alaska. *Condor* 61:233-64.

Poirier, F. E. 1969. Behavioral flexibility and intertroop variation among Nilgiri langurs *(Presbytis johnii)* of South India, *Folia Primatol.,* 11:119-33.

Pollock, J. I. 1975a. Field observations on *Indri indri:* a preliminary report. In *Lemur Biology,* ed. I. Tattersall and R. W. Sussman. Plenum Press, New York and London.

——. 1975b. The social organization and ecology of *Indri indri.* Ph.D. Dissertation, London University.

Post, D. Feeding and ranging behavior of the yellow baboon, *Papio cynocephalus,* Ph.D. Dissertation in prep., Yale University.

Prost, J. H. 1965. A definitional system for the classification of primate locomotion. *Amer. Anthrop.* 67:1198-1214.

Rainey, M. 1970. Aspects of physiological and behavioural temperature regulation in the rock hyrax, *Heterohyrax brucei.* M.A. Thesis, Nairobi College.

Rand, A. L. 1935. On the habits of some Madagascar mammals. *J. Mammal.* 16:89-104.

Richard, A. 1970. A comparative study of the activity patterns and behavior of *Alouatta villosa* and *Ateles geoffroyi. Folia Primatol.* 12:241-63.

Richard, A. F. 1974. Patterns of mating in *Propithecus verreauxi. In Prosimian Biology,* ed. R. D. Martin, G. A. Doyle, and A. C. Walker. Duckworth, London.

——. 1976. Preliminary observations on the birth and development of *Propithecus verreauxi* to the age of six months. *Primates,* 17(3):357-66.

——. in press a. The Feeding Behavior of *Propithecus verreauxi.* In *Primate Feeding*

Behaviour, ed. T. H. Clutton-Brock. Academic Press, London and New York.

———. in press b. Variability in the feeding behavior of a Malagasy prosimian, *Propithecus verreauxi.* In *The Ecology of Arboreal Folivores, Proc. Nat. Zool. Park Symp., No. 1,* ed. G. G. Montgomery. Smithsonian Instituion Press, Washington, D.C.

———, and Heimbuch, R. 1975. An analysis of the social behavior of three groups of *Propithecus verreauxi.* In *Lemur Biology,* ed. I. Tattersall and R. W. Sussman. Plenum Press, New York and London.

———, and Sussman, R. W. 1975. Future of the Malagasy lemurs: Conservation or extinction. Pp. 335-50 in *Lemur Biology,* ed. I. Tattersall and R. W. Sussman. Plenum Press, New York and London.

Richards, P. W. 1966. *The tropical rain forest.* Cambridge University Press.

Ripley, S. 1967. Intertroop encounters among Ceylon gray langurs *(Presbytis entellus).* In *Social communication among primates,* ed. S. A. Altmann. Chicago University Press.

———. 1970. Leaves and leaf-monkeys: The social organization of foraging in gray langurs *Presbytis entellus thersites.* In *Old World monkeys: Evolution, systematics, and behavior,* ed. J. R. Napier and P. H. Napier. Academic Press, New York.

Rodman, P. S. in press. Diets, densities, and distributions of Bornean Primates. In *The Ecology of Arboreal Folivores, Proc. Nat. Zool Park Symp., No. 1,* ed. G. G. Montgomery. Smithsonian Institution Press, Washington, D.C.

Rose, M. D. in press a. Feeding and associated positional behavior of black and white colobus monkeys *(Colobus guereza).* In *The Ecology of Arboreal Folivores, Proc. Nat. Zool. Park Symp., No. 1,* ed. G. G. Montgomery. Smithsonian Institution Press, Washington, D.C.

———. in press b. Positional behavior of olive baboons *(Papio anubis)* and its relationship to maintenance and social activities. *Primates* 18, 1976.

Rowell, T. E. 1966. Forest-living baboons in Uganda. *J. Zool. Lond.* 149:344-64.

———. 1969. Long-term changes in a population of Ugandan baboons. *Folia Primatol* 11:241-54.

Rudran, R. in press. Intergroup dietary differences and folivorous tendencies of two groups of blue monkeys, *Cercopithecus mitis.* In *The Ecology of Arboreal Folivores, Proc. Nat. Zool. Park symp., No. 1,* ed. G. G. Montgomery. Smithsonian Institution Press, Washington, D.C.

Sade, D. S. 1972. A longitudinal study of social behavior of rhesus monkeys. In *The Functional and Evolutionary Biology of Primates.* Aldine Atherton, Chicago.

Schaller, G. B. 1963. *The mountain gorilla.* Chicago University Press.

Schimper, A. F. W. 1903. *Plant-geography upon a physiological basis.* Clarendon Press, Oxford.

Schoener, T. W. 1971. Theory of feeding strategies. *Ann. Rev. Ecol. and Syst.* 2:369-404.

———. 1974. Resource partitioning in ecology communities. *Science* 185:27-39.

Shaw, G. A. 1879. A few notes upon four species of lemurs, specimens of which were brought alive to England in 1878. *Proc. Zool. Soc. Lond.,* pp. 132-36.

Simpson, G. G. 1945. The principles of classification and a classification of mammals. *Bull. Am. Mus. Nat. Hist.* 85:1-350.

Simpson, M. J. A. 1973. The social grooming of male chimpanzees. In *Comparative Ecology and Behaviour of Primates,* ed. R. P. Michael and J. H. Crook. Academic Press, London and New York.

Southwood, T R. E. 1966. *Ecological methods: with particular reference to the study of insect populations.* Chapman and Hall Ltd., London.

Sparks, J. 1967. Allogrooming in Primates: a review. In *Primate ethology,* ed. D. Morris. Weidenfeld and Nicolson, London.

Stoltz, L. P., and Saayman, G. S. 1970. Ecology and behaviour of baboons in the Northern Transvaal. *Annals of Transvaal Mus.* 26:5.

Struhsaker, T. T. 1967. Ecology of vervet monkeys *(Cercopithecus aethiops)* in the Masai-Amboseli Game Reserve, Kenya. *Ecology* 48:891-94.

——. 1969. Correlates of ecology and social organization among African Cercopithecines. *Folia Primatol.* 11:80-118.

——. 1974. Correlates of ranging behaviour in a group of Red Colobus monkeys *(Colobus badius tephrosceles). Amer. Zool.* 14:177-84.

——. 1975. *The red colobus monkey.* Chicago University Press.

——. in press. Interrelations of red colobus monkeys and rainforest trees in the Kibale Forest, Uganda. In *The Ecology of Arboreal Folivores, Proc. Nat. Zool. Park Symp., No. 1,* ed. G. G. Montgomery. Smithsonian Institution Press, Washington, D.C.

——, and Oates, J. 1975. Comparison of the behavior and ecology of Red Colobus and Black-and-White Colobus Monkeys in Uganda: A summary. Pp. 103-23 in *Socioecology and Psychology of Primates,* ed. R. H. Tuttle. Mouton Publishers, The Hague, Paris.

Sugiyama, Y. 1967. Social organization of hanuman langurs. In *Social communication among primates,* ed. S. A. Altmann. Chicago University Press.

Sussman, R. W. 1974. Ecological distinctions in sympatric species of *Lemur.* Pp. 75-108 in *Prosimian Biology,* ed. R. D. Martin, G. A. Doyle, A. C. Walker. Duckworth, London.

——. in press a. Feeding behavior of *Lemur catta* and *Lemur fulvus.* In *Primate Feeding Behavior,* ed. T. H. Clutton-Brock. Academic Press, London and New York.

Suzuki, A. 1969. An ecological study of chimpanzees in a savanna woodland. *Primates* 10:103-48.

Tattersall, I. 1975. Notes on the cranial anatomy of the subfossil Malagasy Lemurs. In *Lemur Biology,* ed. I. Tattersall and R. W. Sussman. Plenum Press, New York and London.

——, and Sussman, R. W., eds. 1975. *Lemur Biology,* Plenum Press, New York and London.

——, and Sussman, R. W. 1975. Observations on the ecology and behavior of the mongoose lemur *Lemur mongoz mongoz* Linnaeus (Primates, Lemuriformes), at Ampijoroa, Madagascar. *Anthrop. Papers of the Amer. Mus. Nat. Hist.* 52(4):193-216.

Thorington, R. W. 1967. Feeding and activity of *Cebus* and *Saimiri* in a Colombian forest. In *Neue Ergebnisse der Primatologie*, ed. D. Stark, R. Schneider, and H. -J. Kuhn, Fischer, Stuttgart.

Treshow, M. 1970. *Environment and plant response*. McGraw-Hill, New York.

Trivers, R. L. 1972. Parental Investment and Sexual Selection. In *Sexual selection and the descent of man*, ed. B. Campbell, Aldine Publishing Co., Chicago.

Uilenberg, G., Blancou, J., and Andrianjafy, G. in press. Un nouvel hématozoaire d'un lémurien malgache *Babesia Propitheci* sp. n. (Babesiidae, Sporozoa). *Ann. Parasit. hum. comp.*

Walker, A. C. 1967. Patterns of extinction among the subfossil Madagascan lemuroids. Pp. 425-32 in *Pleistocene extinctions: the search for a cause*, ed. P. S. Martin and H. E. Wright, Yale University Press, New Haven.

Waser, P. M., and Floody, O. 1975. Ranging patterns of the Mangabey, *Cercocebus albigena*, in the Kibale Forest, Uganda. *Z. Tierpsychol.* 35:85-101.

Wilson, E. O. 1975. *Sociobiology, The New Synthesis*. Belknap Press of Harvard University Press, Cambridge, Mass.

Yoshiba, K. 1968. Local and intertroop variability in ecology and social behaviour of common Indian langurs. In *Primates: Studies in adaptation and variability*, ed. P. C. Jay, Holt, Rinehart and Winston, New York.

Zuckerman, S. 1932. *The social life of monkeys and apes*. K. Paul, Trench, Trubner, London.

Index